Membrane Distillation Process

Membrane Distillation Process

Editor

Alessandra Criscuoli

MDPI • Basel • Beijing • Wuhan • Barcelona • Belgrade • Manchester • Tokyo • Cluj • Tianjin

Editor
Alessandra Criscuoli
Istituto per la Tecnologia delle
Membrane (CNR-ITM)
Consiglio Nazionale delle
Ricerche
Rende (CS)
Italy

Editorial Office
MDPI
St. Alban-Anlage 66
4052 Basel, Switzerland

This is a reprint of articles from the Special Issue published online in the open access journal *Membranes* (ISSN 2077-0375) (available at: www.mdpi.com/journal/membranes/special_issues/membrane_distillation_process).

For citation purposes, cite each article independently as indicated on the article page online and as indicated below:

LastName, A.A.; LastName, B.B.; LastName, C.C. Article Title. *Journal Name* **Year**, *Volume Number*, Page Range.

ISBN 978-3-0365-1211-2 (Hbk)
ISBN 978-3-0365-1210-5 (PDF)

Contents

 membranes MDPI

Editorial
Membrane Distillation Process

Alessandra Criscuoli 

Institute on Membrane Technology (CNR-ITM), via P. Bucci 17/C, 87036 Rende (CS), Italy; a.criscuoli@itm.cnr.it

The water stress that we have been experiencing in the last few years is driving the development of new technologies for the purification and recovery of water. Membrane Distillation (MD) is based on the use of hydrophobic microporous membranes that prevent the passage of aqueous feed as a liquid through the micropores, allowing the transport of water vapor and volatiles only, thanks to a difference in partial pressures established across the membrane. In this way, high-purity distillates can be produced, starting from a variety of aqueous streams, like effluents coming from the textile/agrofood/pharmaceutical industry, olive mill wastewaters, waters contaminated by heavy metals, sea, and brackish waters. Some studies on the application of MD for the purification of radioactive wastewaters and of urine have also been carried out.

With respect to Reverse Osmosis (RO), which is limited by the osmotic pressure and sometimes shows low rejection values for elements like As(III) and Boron, MD is able to produce fresh water from high-concentrated streams and provides 100% theoretical rejections for all non-volatiles present in the aqueous feeds.

Despite these advantages, MD is far from a significant application at industrial scale, due to some pending issues:

- The need to develop membranes with high hydrophobicity and liquid entry pressure values, which can remain stable when treating real streams in long-term runs;
- The need to develop modules with reduced thermal and mass transfer resistances;
- The need to reduce the specific thermal energy consumption.

Research on the above-mentioned points is in progress, including the use of renewable energies to cover the thermal demand of the system and the integration of MD with other membrane units, in order to improve the overall performance of the processes.

The aim of this Special Issue is to provide an overview of the latest results obtained in the field for overcoming MD drawbacks and boosting its implementation at a large scale. Research efforts into the development of new membranes and modules, analysis of the MD performance for bio-feeds at relatively low temperatures, integration of MD with other membrane units and evaluation of MD at pilot scale are presented and discussed.

The mechanical resistance of a PVDF membrane was enhanced by preparing three-layered membranes by electrospinning the PVDF nanofibers on both sides of commercial PES nanofiber-based nonwoven supports [1]. Prepared membranes were first characterized and then tested in Direct Contact Membrane Distillation (DCMD) for the treatment of a 5 M NaCl feed. During 380′ of the test, complete salt rejections were registered, together with trans-membrane fluxes higher than the commercial PVDF membrane, due to the higher porosity of the produced composite membranes.

The mechanical properties and stability during DCMD tests of capillary PP membranes containing talc was investigated by Gryta [2]. Both PP and PP-containing-talc membranes were characterised and tested. The presence of talc led to an increase in porosity and enhanced the mechanical properties due to a disorder of the linear structure of PP. Distillation experiments were carried out on a 1 g/L NaCl feed for 350 h and both membranes led to constant flux and complete rejections. However, process efficiencies higher more than 10% were obtained with the talc addition. After tests, no talc leaching from the membrane was observed.

Citation: Criscuoli, A. Membrane Distillation Process. *Membranes* **2021**, *11*, 144. https://doi.org/10.3390/membranes11020144

Received: 9 February 2021
Accepted: 17 February 2021
Published: 18 February 2021

Composite membranes with hydrophobic coatings on hydrophilic supports were developed to enhance the trans-membrane flux in direct contact membrane distillation tests on salty solutions [3]. The use of hydrophilic porous polyethersulfone as a support led to a reduction in the gas gap for the water vapor transport and, therefore, at 70 °C, fluxes ranging from 27 to 18 kg/m^{2}h were obtained for feed of 1 wt% and 10 wt% NaCl, respectively. The salt passage in the permeate was quite low for the more diluted feed (1 wt% NaCl), while it increased at higher salt content, indicating the need for longer time and/or a higher power supply for the coating step.

A crucial aspect in the design of membrane distillation modules is the choice of spacers, as they are a means to increase mixing of the involved streams, with improvements in heat transfer, but also a cause of pressure drops and reduced membrane area available for evaporation (when the spacers contact the membrane surface). Novel protopypes of overlapped spacers were designed and their performance was investigated by CFD simulations [4]. Spacers differed in terms of geometrical configuration (intrinsic angle between filaments and flow attack angle values). The simulation led to the calculation of the heat transfer coefficient as a function of the Reynolds number (varied from 200 to 800) and to the identification of the optimal geometrical configuration.

An important parameter to improve in view of a large-scale implementation of MD, is the water recovery ratio achievable in a single pass. In fact, low recovery ratios per pass are usually obtained in MD modules, and feed re-circulation or more MD modules working in series are needed to produce a reasonable quantity of distillate and to concentrate the feed up to desired values. A new module design, called the feed gap air gap MD (FGAGMD), was produced in plate and frame geometry, and tested with both tap water and NaCl feeds (117–214 g NaCl/kg) [5]. In the module, the feed was heated through a polymer film in contact with a heating solution, while the distillate condensed on another polymer film in contact with the coolant. If compared with an AGMD spiral wound module, the new module led to a significant improvement in the recovery ratio, with the other performance indicators being similar.

Most of the modules used in MD have an external housing and are characterized by the re-circulation of the feed to be treated. This means the need for a pump for re-circulation and a heater for warming up the stream back to the operating temperature. Different designs of submerged capillary modules were realized and tested by immersing the capillary fibers directly into the hot feed tank (1 g/L NaCl) while recirculating the distillate inside their lumen [6]. Experiments were carried out by changing the type of membrane (material, pore size, inner diameter, thickness) as well as its length and the module packing density. A process efficiency similar to conventional capillary modules was obtained.

The investigation of the MD performance for treating bio-feeds is an important step to evaluate its potential in the field, where relatively low temperatures must be used. Bioethanol is generated, starting from renewable materials through saccharification and fermentation processes, and its production is linked to the sugar content in the prehydrolyzates. Sweep Gas Membrane Distillation (SGMD) was used to concentrate glucose syrup and the parameters that mostly affected the permeate flux were identified [7]. Feed temperature and sweep gas flow rate had the most significant impact; however, in all tests, 99% of glucose rejection was registered, confirming the efficiency of the process for obtaining concentrated glucose streams.

The potential of DCMD to treat at low-temperature feeds containing urea (urine and human plasma ultrafiltrate) was also presented [8]. Complete rejections for urea were registered, allowing the production of concentrate sources for urea/nutrients recovery, together with purified water as a distillate. As a case study, modules of different sizes and properties were tested to improve the permeate production during the treatment of the human plasma ultrafiltrate, and inputs for further research were identified.

The integration of different membrane units is often a means to improve the overall performance of the process, overcoming the limits of single units. An integrated membrane system consisting of Ultrafiltration (UF), Reverse Osmosis (RO) and DCMD was studied for the treatment of the wastewater coming from a flue gas desulfurization (FGD) plant [9]. A total water recovery of 94% was obtained, with a reduction in the volume of wastewater to be disposed. Moreover, the MD permeate stream was suitable to be reused in the power plant.

The evaluation of MD at pilot scale is a crucial aspect for its implementation at the industrial level. SGMD was investigated and found to concentrate (from 30% to 50%) in a commercial module of 1.2 m^2 membrane area aqueous solutions of 1,3-dimethyl-2-imidazolidinone (DMI), a solvent used in different fields, like organic reactions and polymer manufacturing [10]. DMI losses were below 1% of the evaporated flux and more than 99.2% of DMI was recovered in the concentrated solution. Based on the obtained results, empirical models were applied to scale-up the process, evidencing the need for improvements for industrial application.

The production of a permeate with a constant quality in time is essential for the large-scale application of MD. Three commercial spiral wound modules with an area ranging from 7.2 to 24 m^2 were used to evaluate the permeate quality during desalination (feeds of 0.6–2.4 M) [11]. During more than 300 h of test, all modules led to a good permeate quality, with rejections higher than 99%. However, a poor permeate quality was obtained when the process was restarted (intermittent operation), probably due to the presence of membrane pinholes. Further investigations are needed to better understand the behaviour, especially at high-salinity feeds.

The presented papers clearly show the efforts and trends in Membrane Distillation research and development.

My sincere thanks to all the contributors, who make this Special Issue successful.

Funding: This research received no external funding.

Conflicts of Interest: The author declares no conflict of interest.

References

1. Al-Furaiji, M.; Arena, J.T.; Ren, J.; Benes, N.; Nijmeijer, A.; McCutcheon, J.R. Triple-Layer Nanofiber Membranes for Treating High Salinity Brines Using Direct Contact Membrane Distillation. *Membranes* **2019**, *9*, 60. [CrossRef] [PubMed]
2. Gryta, M. The Influence of Talc Addition on the Performance of Polypropylene Membranes Formed by TIPS Method. *Membranes* **2019**, *9*, 63. [CrossRef] [PubMed]
3. Sharma, A.K.; Juelfs, A.; Colling, C.; Sharma, S.; Conover, S.P.; Puranik, A.A.; Chau, J.; Rodrigues, L.; Sirkar, K.K. Porous hydrophobic-hydrophilic composite hollow fiber and flat membrane prepared by plasma polymerization for direct contact membrane distillation. *Membranes* **2021**, *11*, 120. [CrossRef] [PubMed]
4. La Cerva, M.; Cipollina, A.; Ciofalo, M.; Albeirutty, M.; Turkmen, N.; Bouguecha, S.; Micale, G. CFD Investigation of Spacer-Filled Channels for Membrane Distillation. *Membranes* **2019**, *9*, 91. [CrossRef] [PubMed]
5. Schwantes, R.; Seger, J.; Bauer, L.; Winter, D.; Hogen, T.; Koschikowski, J.; GeiBen, S.-U. Characterization and Assessment of a Novel Plate and Frame MD Module for Single Pass Wastewater Concentration-FEED Gap Air Gap Membrane Distillation. *Membranes* **2019**, *9*, 118. [CrossRef] [PubMed]
6. Gryta, M. The Application of Submerged Modules for Membrane Distillation. *Membranes* **2020**, *10*, 25. [CrossRef] [PubMed]
7. Shirazi, M.M.A.; Kargari, A. Concentrating of Sugar Syrup in Bioethanol Production Using Sweeping Gas Membrane Distillation. *Membranes* **2019**, *9*, 59. [CrossRef] [PubMed]
8. Criscuoli, A.; Capuano, A.; Andreucci, M.; Drioli, E. Low-Temperature Direct Contact Membrane Distillation for the Treatment of Aqueous Solutions Containing Urea. *Membranes* **2020**, *10*, 176. [CrossRef] [PubMed]
9. Conidi, C.; Macedonio, F.; Ali, A.; Cassano, A.; Criscuoli, A.; Argurio, P.; Drioli, E. Treatment of Flue Gas Desulfurization Wastewater by an Integrated Membrane-Based Process for Approaching Zero Liquid Discharge. *Membranes* **2018**, *8*, 117. [CrossRef] [PubMed]
10. Abejon, R.; Saidani, H.; Deratani, A.; Richard, C.; Sanchez-Marcano, J. Concentration of 1,3-dimethyl-2-imidazolidinone in Aqueous Solutions by Sweeping Gas Membrane Distillation: From Bench to Industrial Scale. *Membranes* **2019**, *9*, 158. [CrossRef] [PubMed]
11. Ruiz-Aguirre, A.; Andrés-Manas, A.; Zaragoza, G. Evaluation of Permeate Quality in Pilot Scale Membrane Distillation Systems. *Membranes* **2019**, *9*, 69. [CrossRef] [PubMed]

Article

Triple-Layer Nanofiber Membranes for Treating High Salinity Brines Using Direct Contact Membrane Distillation

Mustafa Al-Furaiji [1,2], Jason T. Arena [3], Jian Ren [3], Nieck Benes [4], Arian Nijmeijer [2] and Jeffrey R. McCutcheon [3,*]

1 Environment and Water Directorate, Ministry of Science and Technology, Baghdad 10001, Iraq; alfuraiji79@gmail.com
2 Inorganic Membranes, Faculty of Science and Technology, Mesa+ Institute for Nanotechnology, University of Twente, P.O. Box 217, 7500 AE Enschede, The Netherlands; a.nijmeijer@utwente.nl
3 Department of Chemical and Biomolecular Engineering, University of Connecticut, 191 Auditorium Rd. Unit 3222, Storrs, CT 06269-3222, USA; jta08003@uconn.edu (J.T.A.); jian.ren@uconn.edu (J.R.)
4 Films in Fluids, Faculty of Science and Technology, Mesa+ Institute for Nanotechnology, University of Twente, P.O. Box 217, 7500 AE Enschede, The Netherlands; n.e.benes@utwente.nl
* Correspondence: jeffrey.mccutcheon@uconn.edu; Tel.: +1-860-486-4601

Received: 29 March 2019; Accepted: 25 April 2019; Published: 6 May 2019

Abstract: A composite, three-layered membrane for membrane distillation was prepared from electrospun polyvinylidene fluoride (PVDF) nanofibers supported by commercial polyethersulfone (PES) nanofiber based nonwoven from E.I. duPont de Nemours company (DuPont). The membranes were tested in direct contact membrane distillation (DCMD) using a 5.0 M sodium chloride brine as a feed solution. The triple layer membrane combines the hydrophobicity of PVDF and the robustness of the PES. The triple layer membrane demonstrated excellent performance in DCMD (i.e., relatively high water flux compared to the commercial PVDF membrane and a complete salt rejection of the brine) with mechanical properties imparted by the PES layer. This work is the first to demonstrate the use of a commercially produced nanofiber nonwoven for membrane distillation.

Keywords: membrane distillation; triple layer composite membrane; highly concentrated solutions; PVDF; PES

1. Introduction

Membrane distillation is a thermally driven separation process in which only water vapor molecules transfer across microporous hydrophobic membranes. In direct contact membrane distillation (DCMD), a hot feed solution flows on one side of the hydrophobic membrane while a cold solution (normally DI water) flows on the other side generating the driving force (i.e., vapor pressure) across the membrane. DCMD has been considered as an alternative to traditional distillation to treat highly concentrated solutions and brines due to its low sensitivity towards salinity of the feed solution [1]. These brines may include reverse osmosis reject [2], produced water [3], forward osmosis draw solutions [4] and hypersaline lakes [5]. DCMD among other MD configurations has the advantages of simple design and ease of operation. However, the high heat loss by conduction across the membrane can reduce the performance of the process.

A suitable membrane for membrane distillation should have high porosity, low tortuosity, and high thermal, mechanical and chemical stability [6]. Conventional commercial membranes that are typically prepared by phase inversion have been investigated for the treatment of highly concentrated solutions by membrane distillation [1,7].

Electrospinning is a versatile technique that can be used to prepare nanofiber membranes at desired properties. Recently, nanofiber membranes have been considered for MD [8–10] because of their intrinsically high porosity and low tortuosity [11]. These properties facilitate vapor transport through a membrane yet often lead to a lack of mechanical strength which limit their usefulness to more conventional cast membranes [12].

Adding additional layers to a membrane, such as a supporting scrim, is common with membranes intended for pressurized membrane processes (such as reverse osmosis). Layering MD membranes for the purpose of increasing mechanical integrity has been less common. Prince and co-workers, for example, have studied multi-layered membranes for MD. One of their membranes consists of a top selective layer of electrospun PVDF nanofibers, a middle layer that is made of PVDF formed by immersion precipitation, and a bottom hydrophilic PET support layer [13]. Another three-layered membrane consists of a PVDF cast layer sandwiched between two PVDF nanofibers layers, one of which is modified to be hydrophilic [14]. Both membranes showed improved performance in MD over the nanofiber single layer PVDF membrane due to the increased liquid entry pressure of the triple-layer membrane. We borrow from these multi-layer approaches to examine the use of a commercially available nanofibrous nonwoven material from DuPont (DuPont PES) for membrane distillation. Using the DuPont PES, we apply PVDF nanofibers to both sides of the membrane and measure performance under relevant DCMD conditions. PVDF polymer was chosen because of its hydrophobicity and ease of preparation via electrospinning process. The resulting membrane is entirely nanofibrous, thereby retaining high porosity while garnering strength from the PES and hydrophobicity of the PVDF nanofibers on both sides.

2. Materials and Methods

2.1. Materials

Sodium chloride (NaCl, crystalline, certified ACS) was obtained from Fisher Scientific (Pittsburgh, PA, USA). Acetone and dimethylformamide (DMF) were purchased from Sigma Aldrich (St. Louis, MO, USA). The polymer used in this study was Solef L3 polyvinylidene fluoride generously provided by Solvay Specialty Polymers (Alpharetta, GA, USA). PVDF-HVHP membrane with 0.45 μm pore size was purchased from Millipore and used as a control. This membrane has a porosity of 75% and a thickness of 105 μm. SEM image and other characteristics of the PVDF-HVHP membrane are given in [15]. Water used in this study was ultrapure Milli-Q (18.2 MΩ) water produced by a Millipore Integral 10 water system, (Millipore Corporation, Billerica, MA, USA).

2.2. Membrane Fabrication

To prepare the triple layer MD membrane, a solution of 11.1% PVDF dissolved in 4:1 DMF and acetone was electrospun directly onto a commercially produced nanofiber based nonwoven support from E.I. duPont de Nemours company (Wilmington, DE, USA). DMF alone is a good solvent for the PVDF. However, adding acetone to the solvent plays a key role in accelerating solvent evaporation due to its high vapor pressure leading to the formation of the electrospun nanofibers [16].

This nonwoven support was made of PES fibers spun using a proprietary technique. This material was selected as the support because it can be produced at scale, it has substantially higher strength than lab-made electrospun nanofibers, and the PES and PVDF have a common solvent in DMF. This last feature is important since the PVDF nanofibers will adhere to the PES support as residual solvent promotes fiber bonding. The finished structure consisted of a layer of PVDF fibers on both sides of the PES.

The PVDF fibers were prepared using a lab-scale electrospinning system described in our previous studies [11,17]. The PVDF solution was dispensed through a 20 gauge blunt needle at 5 mL·h-1 with an

18kV potential and a tip-to-target distance of 26 cm. The needle oscillates side-to-side and the collector surface is a grounded rotating drum that was wrapped with the PES support. Spinning was done at room temperature and 65–75% relative humidity (controlled by compressed air). Three thicknesses were prepared: 85 microns, 115 microns, and 145 microns. The thickness of the membrane was controlled by the volume of the PVDF polymeric solution that was spun on both sides of the DuPont PES. As a control membrane, a mat consisting of only PVDF nanofibers with thickness of 85 microns was spun at 18 kV and 65–75% relative humidity.

2.3. DCMD Performance Tests

The DCMD tests were carried out using a customized flat sheet membrane contactor system. The installation consists of two tanks: one for the feed solution and the other for permeate. A warm (50 °C) 5M NaCl solution was used as a feed in all experiment while DI water (20 °C) was used as permeate. Feed solution and permeate were pumped to the membrane cell using variable speed gear pumps (Cole-Parmer, Vernon Hills, IL, USA). A schematic diagram of the system is given in Figure 1. The flowrates of the feed and permeate were kept at 0.4 L/min. Four thermocouples, connected to a 4-channel temperature controller (Sper Scientific Direct, Scottsdale, AZ, USA) were used to measure the temperature at the inlets and the outlets of the membrane cell.

The water flux was calculated based on the weight change of the collected permeate over the experiment time (6 h). The salt rejection R was calculated from the following equation:

$$R = \left(1 - \frac{C_p}{C_{f,i}}\right) \times 100\% \quad (1)$$

where $C_{f,i}$ (g/L) is the initial concentration of the feed solution, and C_p (g/L) is the solute permeate concentration calculated from:

$$C_p = \frac{C_{p,f}\left(V_{p,i} + \Delta V\right) - C_{p,i} V_{p,i}}{\Delta V} \quad (2)$$

In this equation $C_{p,i}$, and $C_{p,f}$ are the initial and final solute concentration at the permeate side, $V_{p,i}$ is the initial permeate volume and ΔV is the permeate volume change during the experiment. The concentration of the solute in the permeate tank was calculated from measuring the conductivity of permeate using an YSI 3200 conductivity meter (Xylem Inc., Yellow Springs, OH, USA).

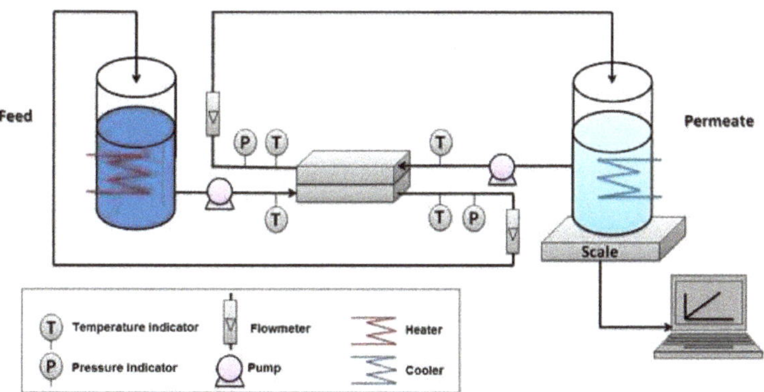

Figure 1. Schematic diagram of the direct contact membrane distillation (DCMD) bench-scale test unit.

2.4. Membrane Characterization

Scanning electron microscopy (SEM) was carried out using cold cathode field emission scanning electron microscope (FESEM, JSM-6335F, JEOL Ltd., Tokyo, Japan) to evaluate fiber size and the surface morphology of the PVDF and PES membranes. Prior to imaging, the samples were sputter-coated with a thin layer of gold. Imaging was done using an accelerating voltage of 20 kV and a current of 1.6 nA. Energy-dispersive X-ray spectroscopy (Thermo Noran System Six EDS, Thermo Electron Scientific, Madison, WI, USA) mapping was conducted using silicon drift detector series EDS (EDAX SDD-EDS). A CAM 101 series contact angle goniometer (KSV Company, Linthicum Heights, MD, USA) was used to measure the contact angles of water drops on the PVDF and the PES nanofibers at five different locations for each sample. The thickness of the different samples was measured at no less than five different locations of each sample using digital micrometer (Mitutoyo, series 293 IP65, Aurora, IL, USA). The mechanical properties of the different membranes were obtained from the tensile tests in air at 25 °C using an Instron microforce tester. A dynamic mechanical analysis (DMA) controlled force module was selected and a minimum of three strips (with a size of 40 mm × 5.5 mm) were tested from each type of membrane.

The porosity of the membranes was estimated using the gravimetric method. The membrane was cut into disks with a diameter of 2.54 cm (1 in) and weighed (W_{dry}). Isopropyl alcohol (IPA) was used as a wetting agent and the membrane weighed after immersed in IPA (W_{wet}). The porosity (ε) was calculated from the following equation [18]:

$$\varepsilon = \frac{\left(\frac{W_{wet}-W_{dry}}{\rho_{IPA}}\right)}{V} \times 100\% \quad (3)$$

where ρ_{IPA} is the density of IPA and V is the total volume of the sample. Each membrane was tested at least three times. The PVDF was assumed not to swell in the presence of IPA.

3. Results and Discussion

3.1. Membrane Characterization

Table 1 shows the properties of the membranes used in this research. The thickness was controlled by the amount of the PVDF polymer that was spun onto the PES layer. The data show that adding the PVDF nanofibers increases the average porosity from 60% to nearly 90%. The added layers of PVDF fibers to both sides of the PES layer increased the thickness of the membrane from 32 to 85, 115 and 145 microns). It can be also seen from Table 1 that the contact angle of the electrospun PVDF membrane is higher than that of the commercial PVDF cast membrane. This is attributed to the difference in the surface roughness as the electrospun nanofibers have ridge-and-valley structure leading to higher roughness compared to the PVDF-HVHP [6].

Table 1. Properties of membranes used in this research.

Membrane	Thickness (micron)	Porosity (%)	Contact Angle (°)
TL-85	85	87.8 ± 1	130 ± 2
TL-115	115	88.6 ± 4	128 ± 5
TL-145	145	88.8 ± 2	129 ± 3
SL-85	85	92 ± 2	132 ± 4
DuPont PES	32	60 ± 1.5	87 ± 1
PVDF-HVHP	105	75	115 ±7

TL: Triple-layer, SL: Single-layer.

The cross-sectional SEM image and the EDS mapping of the triple layer membrane are shown in Figure 2. Here the three distinct layers of the triple-layered membrane can be seen, though it was difficult to section for the SEM imaging. We confirmed the presence of three distinct layers with a sharp interface using EDS mapping. The distribution of the fluoride and the sulfur through the cross-section of the membrane highlight the PVDF and the PES layers respectively. The cross-sectional image also shows the presence of beads in the fibers. Beads are generally not desired in many electrospun nonwovens since they can lead to a weakening of the fiber. In this case, the nanofibrous PES scrim negates this issue. The added roughness may also help prevent wetting of the membrane, but it would certainly be worth investigating further whether or not beads in electrospun fibers have a direct impact on MD performance with nanofiber membranes, which is beyond the scope of this study.

The solvent in the PVDF solution (i.e., DMF/acetone) might result in dissolving or softening some of the nanofibers and enhances binding between the PVDF and PES fibers. The surface SEM images (Figure 3) shows the surface structure of the PVDF and the PES layers. It can be seen that the pore size of the PES layer is smaller than that of the PVDF layer. This is beneficial for MD as it increases the liquid entry pressure of water through the membrane [6]. The contact angle measurements in Table 1 confirm the hydrophobicity of the PVDF nanofibers and the relative hydrophilicity of the PES membrane.

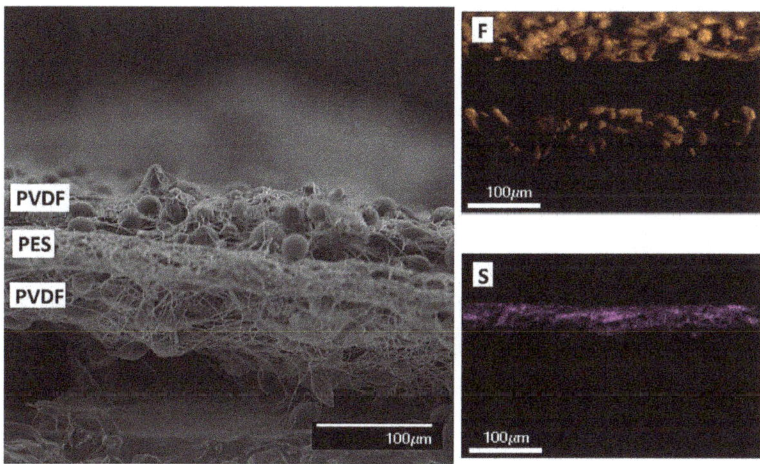

Figure 2. SEM image and energy-dispersive X-ray spectroscopy (EDS) mapping of the cross-section of the TL-115 membrane. Magnification ×230, the yellow color for the fluoride and the purple for the sulfur.

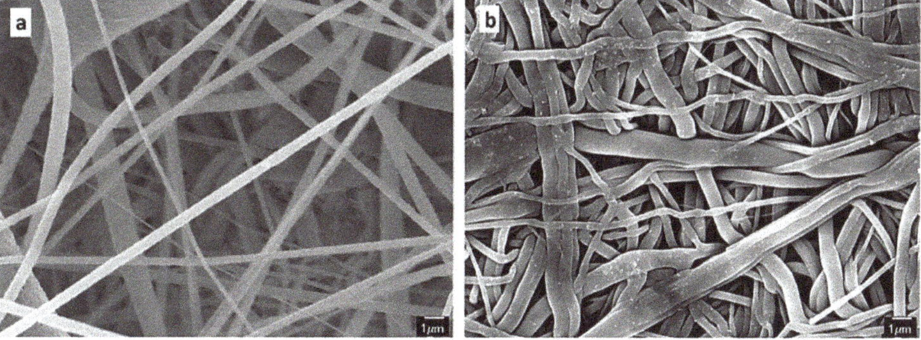

Figure 3. Surface SEM images of the (**a**) polyvinylidene fluoride (PVDF) and (**b**) polyethersulfone (PES) membranes at a magnification of 5000×.

Mechanical properties of the membrane under tension are shown in Figure 4. The PES membrane alone has a tensile strength that far exceeds the PVDF nanofiber nonwovens on their own. Interestingly, after deposition of the PVDF nanofibers onto the PES, the membrane was weaker than the PES membrane alone. Essentially, the tensile test continues until reaching the tensile strength which is the maximum stress that can be achieved before breaking. This method measures the force at break divided by the cross-sectional area of the sample. The PES fibers have higher mechanical strength than the PVDF fibers; adding the PVDF fibers produces a material that has lower mechanical strength than the PES fibers alone. Normalizing for the cross-sectional area gives the force at break, which makes better comparison about the rigidity of the membrane. We present the same data but normalized for the cross-sectional area in Figure 5. Here, the triple layer membrane is shown to withstand a higher force at break and confirms its robustness over the PES or the PVDF nanofibers alone.

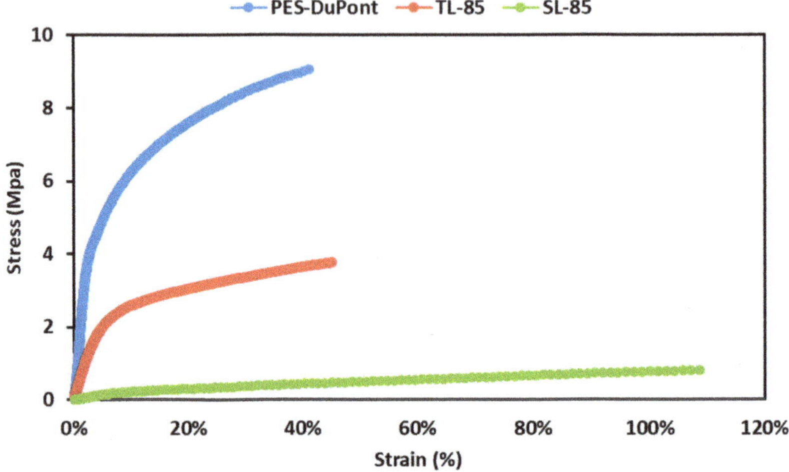

Figure 4. Stress-strain curve of the triple layer membrane, PES and the single layer PVDF membrane.

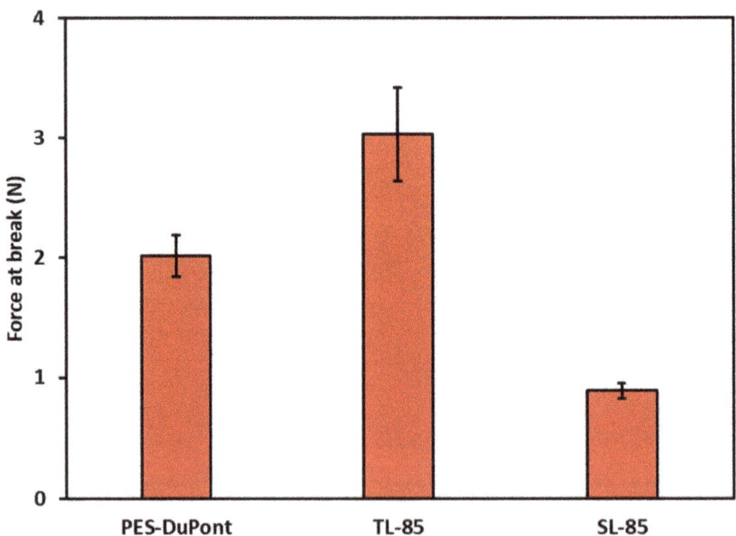

Figure 5. The force at break of the triple layer membrane, PES and the single layer PVDF membrane.

3.2. Membrane Performance by DCMD

DCMD performance was conducted for all the membranes using a 5 M NaCl feed at a 30 °C temperature differential. Testing on the bare PES membrane demonstrated no flux or salt rejection as the membrane wetted out almost immediately. We tested the PES membrane with a single layer of PVDF electrospun nanofibers (fibers on only one side) and the data is shown in Figure 6. While the rejection starts off high (approximately 100%, indicating a properly operating membrane), the rejection noticeably decreases after two hours. This demonstrates that wetting is occurring throughout the two-layer membrane. The large error bars in the rejection are indicative of the variability in the wetting, especially as wetting continues further. Given the relative hydrophilicity of the DuPont PES membrane, it is not surprising to see this variation. Interestingly, as salt rejection decreases, the water flux remains relatively constant. We attribute this to the fact that only small amounts of wetting are occurring as evidenced by the only small reductions in salt rejection. Such a small loss in salt rejection suggests that very little of the membrane is actually wetted from one side to the other. This small amount of salt flux across the membrane does not lead to a substantial loss of vapor pressure driving force, but the loss of rejection is easily detected by even the smallest wetted portion of the membrane since we are using a 5M NaCl feed solution. Similar behavior of dual-layer membranes was reported by Prince and his co-workers [13].

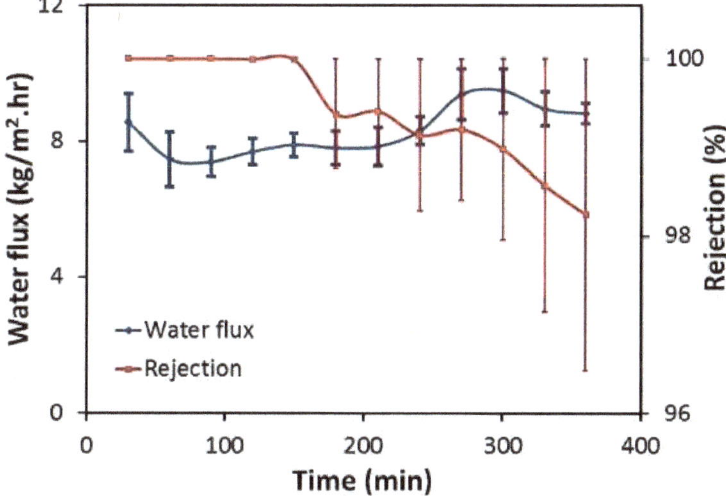

Figure 6. DCMD water flux and rejection for Dual-layer PES-PVDF membrane. Experimental conditions: feed solution: 5 M NaCl, permeate: DI water, feed temperature: 50 °C, permeate temperature: 20 °C, volumetric flow rate of feed and permeate 0.4 L/min.

The performance of all of the triple layer membranes is shown in Figure 7. The fluxes are between 6 and 9 kg/m^2.hr and all membranes exhibit nearly complete salt rejection. The water flux and salt rejection were stable over 6 hours of operating time. In general, our triple-layer membranes exhibited a higher flux than the commercial PVDF-HVHP membrane. The TL-115 membrane showed a water flux twice as high as the water flux of the PVDF-HVHP membrane. This is due to the high porosity of the triple-layer membrane (around 90%) compared to the PVDF-HVHP membrane (75%). Porosity has been reported as the most influencing factor among other membrane properties [19]. Also, the triple-layer membrane showed comparable performance (water flux and salt rejection) compared to the results that were reported in the literature [7,13].

However, it is interesting to note that increasing the membrane thickness from 85 microns to 115 microns gives higher water flux but after increasing the thickness to 145 microns the flux dropped to about the same value. This can be explained by looking into the effect of the added thickness on the competition between heat and mass transport in membrane distillation.

There is a trade-off between mass and heat transfer for the membrane with different thicknesses. A thinner membrane provides a higher rate of mass transport but it also increases the rate of heat transfer by conduction which increases the detrimental effect of temperature polarization [10]. Increasing a membrane's thickness reduces the temperature polarization but also increases the mass transfer resistance. This is valid for a membrane that consists of a single material. The added layer in this research is highly porous. So, increasing the thickness of the membrane (from 32 microns to 85 or 115 microns) will increase the overall porosity and consequently the evaporation area and enhance the mass transfer. However, after adding more thickness (from 115 microns to 145 microns), the negative effect of the large thickness on mass transfer overcomes the positive effect of the high porosity and resulted in lower water flux. We note that in our system, we see the 115 micron thick membrane seems to balance the improved heat transfer resistance with a modest increase in mass transfer resistance and lead to the best water fluxes in this batch of membranes.

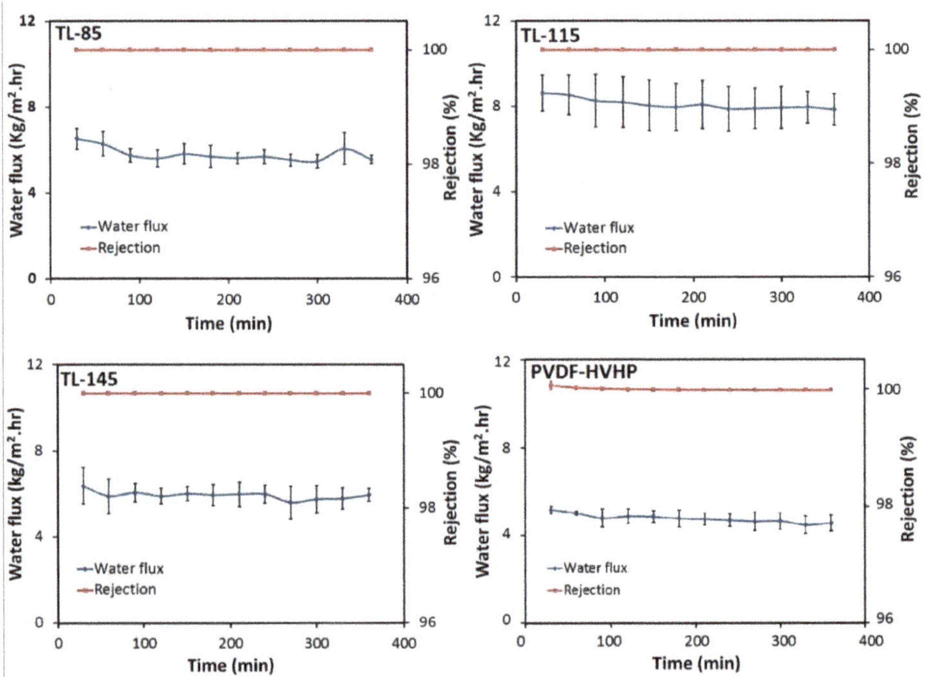

Figure 7. DCMD water flux and rejection for the prepared membranes with different thicknesses. Experimental conditions: feed solution: 5 M NaCl, permeate: DI water, feed temperature: 50 °C, permeate temperature: 20 °C, volumetric flow rate of feed and permeate 0.4 L/min.

4. Conclusions

Triple layer nanofibers membranes with a commercial nanofiber mid-layer were investigated as a mechanically superior alternative to single layer nanofiber membranes. The membranes were comprised of a commercially available PES nanofiber nonwoven, having good mechanical strength, which was used to support a more hydrophobic PVDF electrospun nanofiber. The triple layer membrane showed better performance over both the PES and PVDF single layer membranes. The results also showed that the thickness of the membrane should be chosen carefully due to the interplay between increased heat and mass transfer resistance. When choosing materials for layered membranes, it is critical to understand this interplay and how it can affect performance. It is equally important to consider scalability of the fabrication process to ensure that such membranes have an opportunity to positively impact commercial desalination processes.

Author Contributions: Conceptualization, M.A.-F., J.T.A., and J.R.M.; methodology, M.A.-F.; investigation, M.A.-F., J.T.A., and J.R.; writing—original draft preparation, M.A.-F.; writing—review and editing, M.A.-F., J.T.A., N.B., A.N. and J.R.M.; supervision, N.B., A.N., and J.R.M.

Funding: This research was funded by the higher committee for education development in Iraq (HCED Iraq) No.1000365.

Acknowledgments: The authors would like to acknowledge HCED Iraq for financial funding. The authors also acknowledge the NWRI-AMTA Fellowship for Membrane Technology and National Science Foundation GK-12 Program, which provided support for Jason T. Arena. We also thank DuPont for providing the PES nanofibers. The SEM-EDX studies were performed using the facilities in the UConn/FEI Center for Advanced Microscopy and Materials Analysis (CAMMA).

Conflicts of Interest: The authors declare no conflict of interest.

References

1. Li, J.; Guan, Y.; Cheng, F.; Liu, Y. Treatment of high salinity brines by direct contact membrane distillation: Effect of membrane characteristics and salinity. *Chemosphere* **2015**, *140*, 143–149. [CrossRef] [PubMed]
2. Lattemann, S.; Höpner, T. Environmental impact and impact assessment of seawater desalination. *Desalination* **2008**, *220*, 1–15. [CrossRef]
3. Waisi, B.I.H.; Karim, U.F.A.; Augustijn, D.C.M.; Al-Furaiji, M.H.O.; Hulscher, S.J.M.H. A study on the quantities and potential use of produced water in southern Iraq. *Water Sci. Technol. Water Supply* **2015**, *15*, 370–376. [CrossRef]
4. Al-Furaiji, M.; Benes, N.E.; Nijmeijer, A.; McCutcheon, J.R. Use of FO-MD integrated process in the treatment of high salinity oily wastewater. *Ind. Eng. Chem. Res.* **2019**, *58*, 956–962. [CrossRef]
5. Wang, X.; Miller, J.D.; Cheng, F.; Cheng, H. Potash flotation practice for carnallite resources in the Qinghai Province. *PRC, Miner. Eng.* **2014**, *66*, 33–39. [CrossRef]
6. Tijing, L.D.; Choi, J.S.; Lee, S.; Kim, S.H.; Shon, H.K. Recent progress of membrane distillation using electrospun nanofibrous membrane. *J. Membr. Sci.* **2014**, *453*, 435–462. [CrossRef]
7. Guan, Y.; Li, J.; Cheng, F.; Zhao, J.; Wang, X. Influence of salt concentration on DCMD performance for treatment of highly concentrated NaCl, KCl, $MgCl_2$, and $MgSO_4$ solutions. *Desalination* **2015**, *355*, 110–117. [CrossRef]
8. Prince, J.a.; Singh, G.; Rana, D.; Matsuura, T.; Anbharasi, V.; Shanmugasundaram, T.S. Preparation and characterization of highly hydrophobic poly (vinylidene fluoride)–Clay nanocomposite nanofiber membranes (PVDF-clay NNMs) for desalination using direct contact membrane distillation. *J. Membr. Sci.* **2012**, *397–398*, 80–86. [CrossRef]
9. Liao, Y.; Wang, R.; Tian, M.; Qiu, C.; Fane, A.G. Fabrication of polyvinylidene fluoride (PVDF) nanofiber membranes by electro-spinning for direct contact membrane distillation. *J. Membr. Sci.* **2013**, *425–426*, 30–39. [CrossRef]
10. Essalhi, M.; Khayet, M. Self-sustained webs of polyvinylidene fluoride electrospun nanofibers at different electrospinning times: 1. Desalination by direct contact membrane distillation. *J. Membr. Sci.* **2013**, *433*, 167–179. [CrossRef]
11. Bui, N.N.; McCutcheon, J.R. Hydrophilic nanofibers as new supports for thin film composite membranes for engineered osmosis. *Environ. Sci. Technol.* **2013**, *47*, 1761–1769. [CrossRef] [PubMed]
12. Huang, L.; Arena, J.T.; Manickam, S.S.; Jiang, X.; Willis, B.G.; McCutcheon, J.R. Improved mechanical properties and hydrophilicity of electrospun nanofiber membranes for filtration applications by dopamine modification. *J. Membr. Sci.* **2014**, *460*, 241–249. [CrossRef]
13. Prince, J.a.; Anbharasi, V.; Shanmugasundaram, T.S.; Singh, G. Preparation and characterization of novel triple layer hydrophilic–hydrophobic composite membrane for desalination using air gap membrane distillation. *Sep. Purif. Technol.* **2013**, *118*, 598–603. [CrossRef]

14. Prince, J.A.; Rana, D.; Matsuura, T.; Ayyanar, N.; Shanmugasundaram, T.S.; Singh, G. Nanofiber based triple layer hydro-philic/-phobic membrane – A solution for pore wetting in membrane distillation. *Sci. Rep.* **2014**, *4*, 6949. [CrossRef] [PubMed]
15. Al-Furaiji, M.; Benes, N.; Nijmeijer, A.; McCutcheon, J.R. Application of direct contact membrane distillation for treating high salinity solutions: impact of membrane structure and chemistry. *Desalin. Water Treat.* **2018**, *136*, 31–38. [CrossRef]
16. Santoro, S.; Vidorreta, I.; Sebastian, V.; Moro, A.; Coelhoso, I.; Portugal, C.; Lima, J.; Desiderio, G.; Lombardo, G.; Drioli, E.; et al. A non-invasive optical method for mapping temperature polarization in direct contact membrane distillation. *J. Membr. Sci.* **2017**, *536*, 156–166. [CrossRef]
17. Huang, L.; Bui, N.N.; Manickam, S.S.; McCutcheon, J.R. Controlling electrospun nanofiber morphology and mechanical properties using humidity. *J. Polym. Sci. Part B Polym. Phys.* **2011**, *49*, 1734–1744. [CrossRef]
18. Khayet, M.; Matsuura, T. *Membrane Distillation: Principles and Applications*; Elsevier: Amsterdam, The Netherlands, 2011. [CrossRef]
19. Khayet, M.; Imdakm, A.O.; Matsuura, T.; Carlo, M. Transfer Monte Carlo simulation and experimental heat and mass transfer in direct contact membrane distillation. *Int. J. Heat Mass Transfer* **2010**, *53*, 1249–1259. [CrossRef]

Article

The Influence of Talc Addition on the Performance of Polypropylene Membranes Formed by TIPS Method

Marek Gryta

Faculty of Chemical Technology and Engineering, West Pomeranian University of Technology Szczecin, ul. Pułaskiego 10, 70-322 Szczecin, Poland; marek.gryta@zut.edu.pl

Received: 25 March 2019; Accepted: 7 May 2019; Published: 14 May 2019

Abstract: The effect of talc addition on the morphology of capillary membranes formed by a thermally induced phase separation (TIPS) method was investigated in the presented work. The usability of such formed membranes for membrane distillation was evaluated. Two types of commercial capillary polypropylene membranes, fabricated for microfiltration process, were applied in the studies. A linear arrangement of polymer chains was obtained in the walls of membranes formed without a talc addition. In the case of membranes blended with talc, the linear structure was disordered, and a more porous structure was obtained. The changes in morphology enhanced the mechanical properties of blended membranes, and their lower thermal degradation was observed during 350 h of membrane distillation studies. Long-term studies confirmed the stability of talc dispersion in the membrane matrix. A leaching of talc from polypropylene (PP) membranes was not found during the membrane distillation (MD) process.

Keywords: membrane distillation; membrane stability; polypropylene; TIPS; talc

1. Introduction

The appearance of the MD process can be found in the Bodell patent from 1963 [1,2] and despite the intensive studies carried out for more than 50 years, durable MD membranes have not yet been developed. The hydrophobic porous membranes made from polytetrafluoroethylene (PTFE), polyethylene (PE), polyvinylidene fluoride (PVDF) and polypropylene (PP) are frequently used in the studies of the MD process [2–7]. In several works, the commercial membranes made from these polymers for microfiltration (MF) were tested as the MD membranes [3,8–12]. However, with regard to the MD process conditions (e.g., higher feed temperature and non-wetted pores) the membranes fabricated for MF, in the majority of cases, did not fulfil the MD process requirements. Generally, a membrane for the MD process has to simultaneously meet the following requirements [2,12–19]:

- high liquid entry pressure (LEP), which is the minimum transmembrane hydrostatic pressure that is applied on the membrane before a liquid solution penetrates into the pores;
- good thermal stability—up to the boiling temperature of water;
- high chemical resistance to separated solutions;
- high permeability;
- low thermal conductivity;
- narrow pore size distribution.

A simultaneous fulfilment of these conditions is difficult, e.g., high permeate flux can be achieved for thin membranes, but a low thermal conductivity and higher energy efficiency can be obtained when thicker membranes are used, especially for brines desalination [20,21]. Moreover, a further progress of the MD process (industrial implementation) requires a substantial growth in studies on a semi-pilot or pilot scale [18,21,22].

The pioneering activity within a scope of the MD process implementation is performed, among others, by SolarSpring GmbH (Freiburg, Germany), Memsys GmbH (Schwabmünchen, Germany), TNO (Den Haag, The Netherlands), Aquaver (Voorburg, The Netherlands), Scarab Development AB (Stockholm, Sweden), Xzero AB (Stockholm, Sweden), BlueGold Technologies (Largo, FL, USA) and Abengoa Water (Seville, Spain) [18]. The pilot plants constructed by the above-mentioned companies utilize mainly the membranes made from PTFE, PE and PP recently produced for the MF process [18,23]. Although the applied membranes exhibit certain drawbacks, the realized pilot studies have a positive impact on the demand for the MD process and enhance the opportunity for implementation of novel membrane production in future.

An intensive development of the MD process investigation has been observed in recent years, and as a result, the number of publications has grown from 1000 to almost 5000 [16,18]. Several works have reported that the pores wettability can be restricted by application of composite membranes or membranes with modified surface which enhances the resistance to wetting [8,9,16,24–26]. Promising results were obtained for the membranes formed by electrospinning method and the membranes with addition of different fillers, e.g., carbon nanotubes, titanium and silicon dioxide or talc [27–32]. In addition to membranes made of polymers, ceramic membranes are also presented, which create a chance for applications requiring particularly durable membranes [33,34]. Many new types of membranes and methods for their preparation and the possibility of their use in the MD process are described extensively in review articles [8,16,18,34,35].

Nevertheless, so far, no breakthrough results have been obtained, which indicates that durable membranes for the MD process could be difficult to fabricate. One of a reasons for the slow development of MD membranes is the fact that the majority of works present only ways to increase the LEP value or contact angle, but there are no studies demonstrating the durability of new types of membranes during a long-term module exploitation. Moreover, even the membranes with a very good resistance to wetting, can undergo the progressive degradation due to slow changes in the polymer after 100–200 h of the MD process [36]. Moreover, the implementation of new types of membranes for the production is difficult, mainly due to a fact that the MD membranes market is just being created [18]. Moreover, the manufacture of MD membranes on a larger scale is necessary because it enables the performing of pilot studies [18,22]. Demonstrating the possibility of realizing different applications of the MD process on a pilot scale gives the opportunity to attract industrial investors, what is necessary for process implementation.

With regard to the above-mentioned issues, at the initial stage of MD development it is essential to apply the methods already industrially employed for the production of MD membranes. A thermally induced phase separation (TIPS) process is such a method, commonly used to fabricate the membranes from hydrophobic polymers [1,37,38]. In the TIPS method, the polymer with/without fillers is introduced into the mixer, where is mixed with proper amounts of different kinds of oils [37–43]. The polymer granulates are completely melted in the diluents system, and obtain a hot (e.g., 453–493 K) homogenous dope solution that flows through the spinning nozzle into a coagulation bath. Subsequently, the oils are extracted from the membrane matrix.

It has been demonstrated in numerous papers, that the membranes manufactured by the TIPS method have the appropriate properties for the MD process. The membranes made from PP are most often used for MD [10,18,23,37,38,41,43]. Moreover, the application of other polymers, such as PVDF, PE and polyethylene chlorinetrifluoroethylene (ECTFE) for MD membrane preparation via the TIPS method is also possible [39,44,45]. In several works it has been reported that improvement of the properties of MD membranes could be obtained by modification a composition of dope solution and by using different conditions for the membrane formation. However, in the majority of cases, the studied membranes were formed under laboratory conditions. Therefore, certain phenomena taking place during the membranes production on the industrial scale, such as the row arrangement of polymer chains resulting in the formation of linear structure of the membrane matrix, can affect the polymer mechanical properties [46]. The possibilities of preventing the polymer degradation and methods for improvement of the mechanical properties of PP membranes by introducing fillers into the dope solution, are presented in this paper.

The polypropylene capillary membranes represent the membranes manufactured via the TIPS method on an industrial scale for the MF process [10,23]. The flow of melted polypropylene through a spinning nozzle causes a more linear arrangement of the polymer chains. A cooling-down of formed capillary in the coagulation bath proceeds rapidly [38], which favours the freezing of polymer linear structure. A result is a deterioration of tensile strength manifested by a longitudinal creaking of polymer. The addition of crystals nucleus (e.g., inorganic fillers) into a dope solution is one of the methods of disturbance of the linear arrangement of polymer chains [46,47]. It is well-documented in the literature that the blending of polymers with mineral fillers is considered a useful way to improve the mechanical properties and talc is a popular mineral filler used for this purpose [46]. Talc is a cheap filler and more importantly, it has the hydrophobic properties that enable its very good dispersion in polyolefin. For this reason, talc is often used to improve the properties of materials produced from polyethylene or polypropylene [30,47].

In current MD pilot studies, hydrophobic membranes produced for microfiltration are often used. However, the feed temperature in the MD process, as a rule, is higher than that which is applied in the MF process. As a result, the membrane matrix is subjected to a larger thermal expansion during the MD process, what can lead to damages in the membrane structure. In order to enhance the thermal resistance of produced polymeric materials from PP, the polymer is usually blended with different inorganic fillers, such as talc, which exhibits many actions improving the PP properties [29,47]. It was demonstrated that the application of talc enhanced the thermal resistance and tensile strength of PP membranes [29,39,48,49]. However, as the thickness of the film produced from PP decreases, then the beneficial effects of the fillers used, such as talc, are reduced [46]. Moreover, the structure of the membranes (pore walls) has a much smaller thickness compared to the films produced. For this reason, the objective of this work is to determine the effect of talc addition on the properties of PP membranes used in the MD process. Moreover, PP membranes have been shown to undergo a significant transformation, such as surface wettability, during initial 50–100 h of the MD process [10,36,50]. Therefore, the possibility of application of used membranes in this work over significantly longer periods (350 h) was studied.

The membrane matrix durability studies were often omitted in the previous works describing the MD process. One of the main applications of the MD process is water desalination in order to obtain drinking water. Works in which research is undertaken to determining whether the fillers introduced into the membrane matrix or compounds used to modify the membranes surface will not be released into the produced water are scarce [51]. This information is important since many of the components, such as nanomaterials, used to improve the properties of MD membranes, are suspected of being carcinogenic [52,53]. There are well-known works suggesting that a long-term contact with talc may also have a carcinogenic effect [54]. Until now, such an action has been attributed to talcum forming asbestos-moulded structures [54,55]. However, recent studies indicate that the so-far considered safe platelet forms of talc may also have a negative effect on human health [56]. For this reason, it is important to obtain a kind of membrane matrix that would prevent the leaching of fillers. The present study investigated whether there was no leaching of talc from the surface of the membranes during a long-term MD process.

Taking into account a significant impact of production installation design and the conditions of its exploitation on the properties of fabricated membranes, the studies were performed using two types of capillary MF membranes (with/without talc) manufactured in industrial installation under similar conditions of the TIPS process.

2. Materials and Methods

Two types of capillary PP membranes purchased from PolyMem (Warszawa, Poland) were used in the studies. These membranes have the internal diameter of 1.8 mm and the outer diameter of 2.6 mm.

Although the composition of dope solution and the process parameters of membrane production were similar, about 10% of talc (as a nucleation agents) was added to one dope solution. The membranes utilized in this work were designated as PP-N (net polypropylene) and PP-T (polypropylene with talc). The parameters of used membranes were summarized in Table 1. The manufacturer stated that the average pore diameters equal to 0.2 μm were represented in both cases. The membrane porosity was determined using the gravimetric method [57]. In this case, the membrane samples with a length of 25 cm were soaked in isopropanol for 1 h to achieve a pore wetting.

Table 1. The parameters of applied membranes.

Membrane	Internal Diameter [mm]	Wall [mm]	Porosity [%]	Pore Diameter [μm]
PP-N	1.8	0.4	81	0.2
PP-T	1.8	0.4	84	0.2

The MD module configuration, such as submerged modules, was applied for direct contact MD studies with preferred option. Each module was equipped with two capillaries, and the working length of capillaries was 40 cm (membrane area 45 cm^2). The experimental set-up is shown in Figure 1. The submerged MD module was assembled inside the feed tank, and distillate flowed inside the capillaries (linear velocity of 0.55 m/s). The MD installation was operated, at the feed temperature equal to 353 K, for 350 h in a continuous mode (day and night). Since mainly the thermal effects were studied, a dilute NaCl solution (1 g/L) was used as a feed. When certain pores are wetted in the membrane, the feed may leak to the distillate. Therefore, the presence of NaCl increases the electrical conductivity of the distillate, which allowed for assessing the degree of wetting of tested membranes. The distillate was cooled by tap water and its temperature was maintained at 288–295 K. The permeate flux was calculated on a basis of changes in the distillate volume over a studied period of time (20–24 h). The volume of obtained distillate was in the range of 500–1000 mL over this period. Assuming that the volume was measured with an accuracy of 5 mL and the error of the membrane area calculation was 2%, the error of the MD measurements (permeate flux) did not exceed 2–3%.

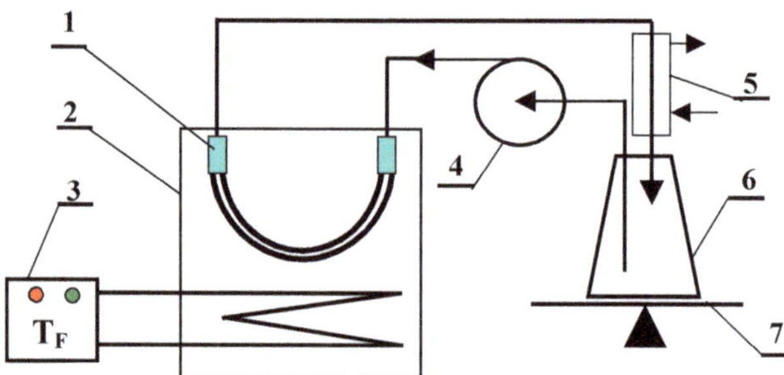

Figure 1. MD experimental set-up. 1—MD module, 2—feed tank, 3—Nũga temperature regulator, 4—peristaltic pump, 5—cooler, 6—distillate tank, 7—balance.

The thermal properties of used PP membranes were determined by differential scanning calorimetry (DSC). These tests were performed by means of a device DSC Q100 (TA Instruments, New Castle, DE, USA), at heating and cooling rate of 10 K/min within the temperature range 200–523 K. The samples were examined in heating-cooling-heating cycles after previous drying.

The mechanical strength and elongation at break of the PP capillary membrane were measured with a tensile tester (3366 Universal Material Testing Machine, Instron, Norwood, MA, USA) according to PN-EN ISO 527-1:1998 method. The specimens of membrane with an initial length of 10 cm were clamped at both ends obtaining the measurement length 3 cm, and pulled in tension at the constant elongation rate of 10 mm/min. In each case, the measurements were performed for ten samples of PP membrane.

The changes of membrane hydrophobicity were determined using a Sigma 701 microbalance (KSV Instrument, Ltd., Espoo, Finland). Based on the Wilhelmy plate method, the dynamic contact angles were measured.

A 6P Ultrameter (Myron L Company, Carlsbad, CA, USA) was applied to measure the values of the electrical conductivity.

The membrane morphology and composition of inorganic fillers were investigated using a Hitachi SU8000 Scanning Electron Microscope (SEM) equipped with Energy-dispersive X-ray Spectrometer (EDS). The specimens for cross-sectional examinations were prepared by fracture of the capillary membranes in liquid nitrogen. Before the SEM examination the membrane samples were sputter coated with chromium (about 3–5 nm) using Q150T ES coater (Quorum Technologies Ltd., Lewes, UK).

In addition to SEM-EDS examinations, the membranes were tested using FTIR, XPS and elemental analysis. This analysis was performed using Elemental Analyzer FLASH 2000 CHNS/O (Thermo Scientific, Waltham, MA, USA), which operates according to the dynamic flash combustion of the sample. For C and H determination the obtained gases were carried by a helium flow to a layer filled with copper, then swept through a GC column, and finally, a Thermal Conductivity Detector (TCD) detected them. In the case of oxygen, the samples were introduced into the pyrolysis chamber via the MAS Plus Autosampler. The reactor contains nickel-coated carbon maintained at 1060 °C. The oxygen in the sample forms carbon monoxide, which is then separated from other products using gas chromatograph and detected by the TCD Detector.

The functional groups presented on the membrane surface were identified by a Fourier transform infrared (FTIR) spectroscopy. The method of Attenuated Total Reflection (ATR) was applied. In the performed studies, a Nicolet 380 FTIR spectrofotometer connected with Smart Orbit diamond ATR instrument (Thermo Electron Corp., Waltham, MA, USA) was used.

Chemical composition of the membrane surfaces was evaluated by X-ray photoelectron spectroscopy (XPS) (Prevac, Rogów, Poland). Prevac electron spectrometer, equipped with an SES 2002 (VG Scienta, Uppsala, Sweden) electron energy analyzer working in a Constant Energy Aperture mode was applied in these studies. The concentration of detected elements (expressed in atomic percents) was calculated in CasaXPS software.

X-ray diffraction (XRD) studies were performed in order to determine a crystal structure of the membranes. In these studies was applied an EMPYREAN diffractometer (PANanalytical, Almelo, The Nederland) using a monochromatized CuKα radiation (35 kV, 30 mA). The obtained peaks that presented the kth Bragg reflection were described by Pseudo-Voigt profile function, which can be assigned a fixed shape of any type between their limiting Gaussian and Lorentzian forms. Based on the Pseudo-Voigt profile, the High Score Plus 3.0 software was applied for estimation of a full-width at half maximum (FWUH) parameter. XRD allowed for indicating a material with different electron density. Assuming an existence of a difference between PP and talc, a trial was undertaken to apply software Easy SAXS 2.0 for determination of talc particle size.

3. Results and Discussion

3.1. SEM Examinations of Membranes

The performed SEM examinations revealed significant differences in the structure of tested membranes, especially that the surfaces of PP-T membranes blended with talc were more porous (Figure 2). These results confirmed that the addition of fillers into a dope solution significantly influences the morphology of membranes formed by the TIPS method [41,42].

In each case studied it was found that both the internal and external surfaces of the capillaries were less porous than a structure formed inside the membrane wall (Figure 2). A reason for different porosity is associated with the leaching of solvent and the rate of cooling, which takes place at a significantly faster rate on the capillary surface than inside the wall [45]. Moreover, in the case of PP-N membrane (without talc) the porosity of the external surface of capillary (Figure 2a) was lower than that observed on the lumen side (Figure 2b). The characteristic row lamellar structure was formed on this side. When a dope solution flowed through the spinning nozzle, the higher shear rate provided the necessary conditions to form highly oriented row nuclei, which induce an epitaxial growth of crystallites [46]. As a result, a linear layer of PP chains oriented parallel to the flow direction of dope solution was formed.

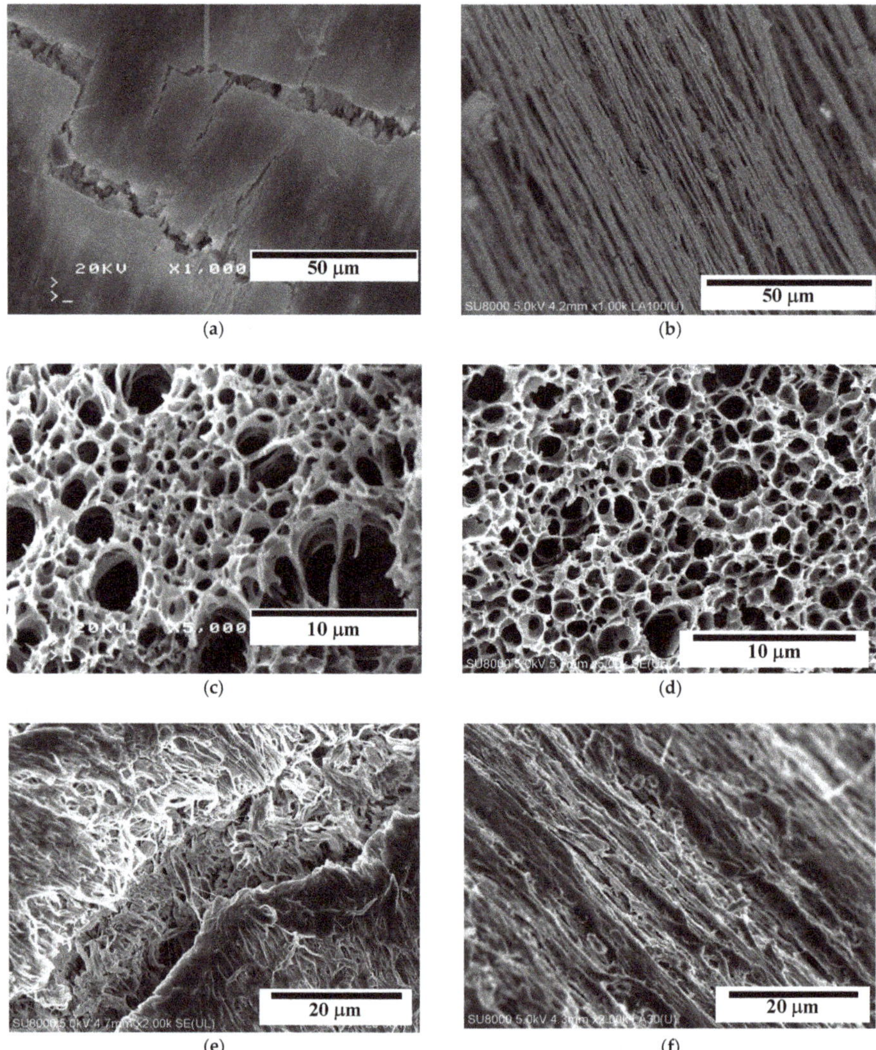

Figure 2. SEM images of new capillary membranes. Membrane PP-N: (**a**) external surface, (**b**) internal surface, (**c**) cross-section. Membrane PP-T (with talc): (**d**) cross-section, (**e**) external surface, (**f**) internal surface.

In the case of PP-T membranes blended with talc, the linear structure was disordered, and a more porous surface was obtained (Figure 2e,f). The final morphology depends on the relaxation behavior of the PP chains during the cooling and solidification process of casting dope [46]. Long polymer chains with a longer relaxation time did not have sufficient time to relax and would remain in the stretched state. The talc addition increases the number of nuclei; therefore, the crystallization rate of PP is significantly accelerated [29,39]. This meant that long polymer chains (linear structure) were not formed or/and that the relaxation time was shorter, and that the PP chains were able to quickly relax back to the coiled state.

Inside the membrane wall the cooling and solidification processes proceed at a significantly slower rate, thus, the polymer chains have sufficient time to relax, and the porous structure without linear orientation can be formed [46]. The SEM examinations revealed that a sponge-like structure was formed in both cases (Figure 2c,d). The addition of talc meant that the pores in the PP-T membrane were more spherical and the smaller pores with the dimension of pore cells in the range of 1–2 μm were formed. Definitely larger differences in the pore size were exhibited in the PP-N membrane. Besides the pores with dimension of 1–2 μm, numerous pores with the size of 3–7 μm also occurred (Figure 2c). Such large pores may facilitate the wetting of PP-N membranes during the MD process.

The SEM examinations of PP-T membrane did not reveal the presence of larger talc particles (i.e., size above 0.1 μm) in the membrane matrix and on the capillary surfaces (Figure 2c–f). The presence of talc particles with a significant size (e.g., 5–10 μm) was only found in a few places of the examined cross-section (Figure 3a). However, the SEM-EDS examinations showing a good dispersion of small talc particles (chemical formula $Mg_3Si_4O_{10}(OH)_2$) in the membrane matrix (Figure 3b–d). This indicated that a dominant fraction of added filler (talc) had a smaller size than that which was observed in Figure 3a.

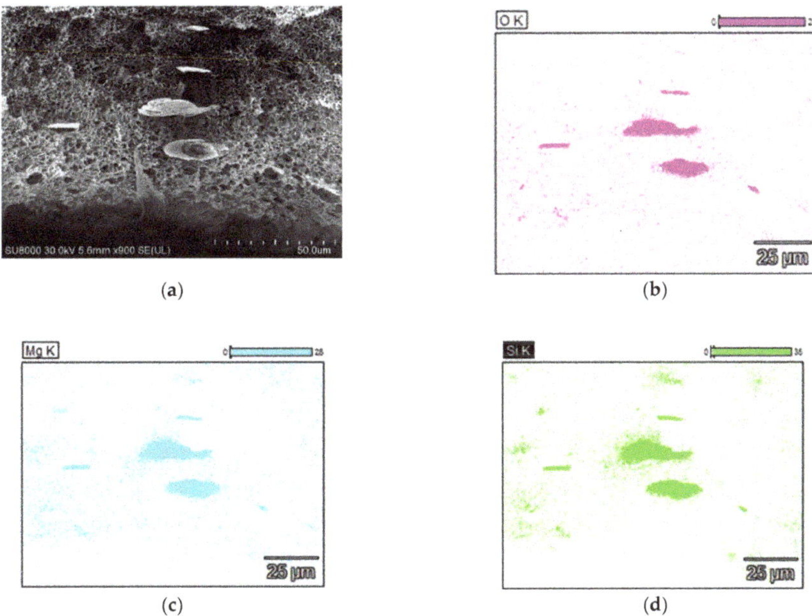

(a) (b) (c) (d)

Figure 3. SEM-EDS analysis of PP-T membrane cross-section (**a**). Elements dispersion: (**b**) oxygen, (**c**) magnesium, (**d**) silicon.

The SEM observations revealed that the talc used for membranes production had a plate structure (Figure 4). The point analysis of the composition made by the SEM-EDS method showed that it contained 46.1% O, 13.1% Mg, 16.4% Si and 24.4% C. Carbon is not a component of talc, but in the case of particles of size about 2 μm (Figure 4) the EDS analysis also includes polypropylene surrounding such a small particle.

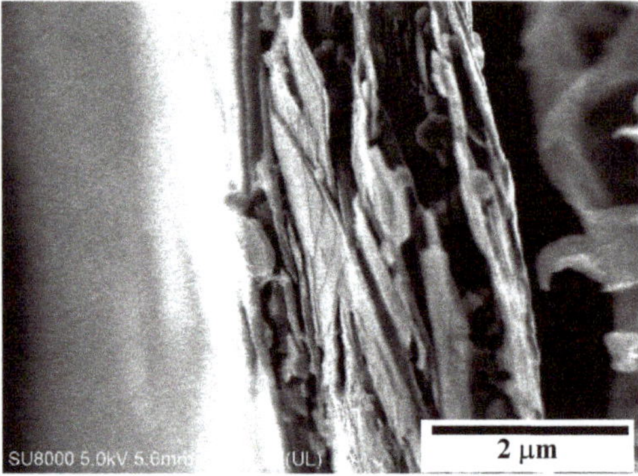

Figure 4. SEM image of talc particle observed inside the PP-T membrane cross-section.

The SEM observations of PP-T membrane samples conducted under high magnification (50–100 k) enabled us to conclude that there was no talc particle size above 50 nm on the surface of the membrane matrix (Figure 5). On the other hand, such large magnifications allowed for observing the formation of shish-kebab crystal structures (Figure 5b), which is characteristic of polypropylene and which can be prepared in polymeric materials by adding nanoparticles or fibres [58]. Such forms were not found when observing PP-N membrane samples. This confirms the beneficial effect of talc as nuclei accelerating the PP crystallizations.

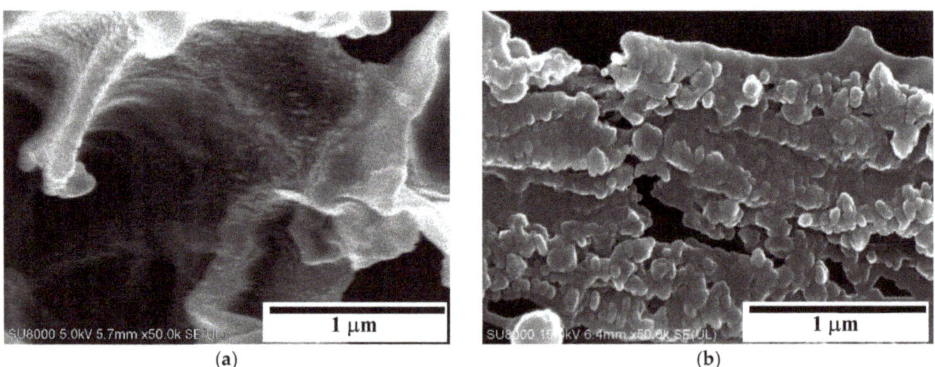

Figure 5. SEM image of the PP-T membrane cross-section. Magnification 50,000×. (**a**) surface inside the pore; (**b**) shish-kebab crystal structures.

An attempt was made to determine the size of talc particles based on XRD measurements. A particle size analysis conducted with the use of Easy SAXS software showed most frequent radius equal to 68.9 nm, and the obtained value of relative standard deviation was 44.4%. Although this result does not contradict the conclusions from SEM observations, it can be burdened with a large method error. The main reason for this is a very low absorption factor of sample (1.313), which may indicate that PP is too light a material (low electron density) for this method. Moreover, the membrane matrix also contained talc particles with sizes over 1000 nm, which are too large for this method and thus could distort the measurement results.

3.2. Long-Term MD Studies

The MD process was carried out using the submerged modules located in the feed tank, and the distillate flows on the lumen side. In the performed MD studies, the feed temperature amounted to 353 K. The application of such temperature or higher, allows for significantly increasing the efficiency of the MD process [10], but disadvantageously accelerates the polymer degradation process. In our case, a high feed temperature applied in the studies carried out for 350 h should facilitate the evaluation of effectiveness of talc addition to the MD membranes. The membrane degradation causes the degree of PP surface oxidation (hydrophilization) to increase and the formation of cracks in the capillary wall can be observed [46]. The hydrophilization of membranes causes a systematic decline of the module productivity and the values of electrical conductivity of obtained distillate are increased [2,9,16,22].

The changes of the permeate flux and electrical conductivity of distillate during long-term MD studies are presented in Figure 6. A comparison of experimental results indicated that the obtained permeate flux for each tested membrane was practically constant during the MD study period (350 h). Small flux discrepancies resulted from the changes in cooling water temperature (Figure 1), because for reasons of safety, the distillate side thermostat was abandoned (additional electric heater). The largest productivity, at a level of 10.4 L/m^{2}h was obtained for the PP-T membrane and a slightly smaller result (9.5 L/m^{2}h) for PP-N membrane. The observed stability of the permeate flux confirms that the process of membrane wettability is restricted.

A higher permeate flux obtained for PP-T membranes results from a larger porosity of these membranes especially for the external surface of capillaries, as was demonstrated by the SEM examinations (Figure 2). These results are also in agreement with those obtained in the previous work, in which the permeate flux was lower by more than 10%, due to the decrease in the surface porosity of used membranes (as in the case of PP-N) [59]. Such a small decrease in efficiency was also obtained in other works wherein the strength of MD membranes was increased by applying an additional surface layer. Such a layer allows for significantly increasing the value of LEP and the contact angle of the membrane surface [24,25,35]. However, this method also increases the surfaces for heterogeneous crystallization. Therefore, to assess the effectiveness of using the membranes for pilot-plant studies, long-term MD studies with intensive scaling should be carried-out.

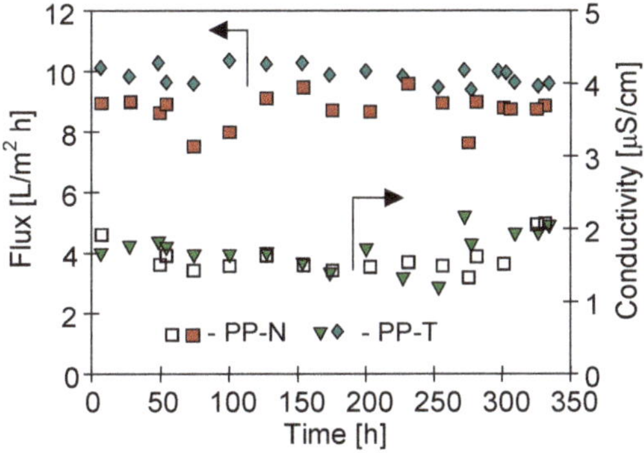

Figure 6. The changes of permeate flux and distillate conductivity during MD process. T_F = 353 K.

The addition of inorganic fillers can accelerate the membrane wetting [60]; therefore, a growth of the electrical conductivity of the MD distillate can be observed. However, the talc is regarded as a hydrophobic material [29,30], thus, it should not facilitate the penetration of water into the pores. This conclusion was supported by a low value of the electrical conductivity of the obtained distillate. The conductivity values at a level of 2 µS/cm were stable for distillate obtained from each module (Figure 6). Taking into account the conductivity of the feed exceeding 2000 µS/cm, it can be concluded that the degree of separation obtained for each of the tested membranes was close to 100%. Such a result and the stable permeate flux confirms that the used PP membranes exhibited a good resistance for wettability.

The membrane wettability is characterized by a value of the contact angle. The dynamic measurement of the contact angle was carried out using the Wilhelmy plate method. The initial values obtained for new membranes were 99° and 104° for PP-N and PP-T membranes, respectively. A morphology of the membrane surface is strongly affected by the value of the contact angle during the first immersion of membrane samples, however, after 20–30 repetitions of immersion–emersion cycles the contact angle values were stabilized, and the values at a level of 82° (PP-N) and 86° (PP-T) were obtained (Figure 7). Similar values of the contact angle were obtained: for PP-N 73.2° and 74.6° for PP-T membranes collected from modules after 350 h of MD process; it can thus be concluded that the addition of talc increased, in a slight degree, the resistance of tested PP-T membranes to wetting. This was mainly due to the fact that the addition of talc to a dope solution allowed for obtaining the pores of similar sizes and eliminating the formation of very large pores, which are easier to wet.

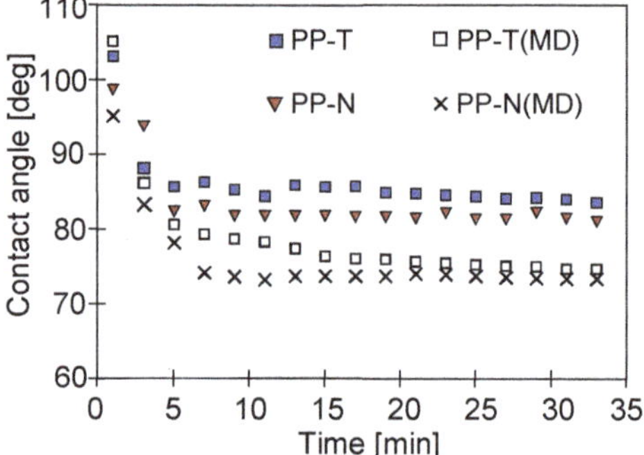

Figure 7. The changes of the contact angle (advancing) during continuous measurements of dynamic contact angle using the Wilhelmy plate method. Samples of new membranes and membranes collected from modules after 350 h of MD process.

The addition of inorganic fillers into the polymeric membrane matrix can enhance the values of thermal conductivity and as a result; the heat losses will be higher in the MD process. However, the results presented in Figure 8 indicate that a higher thermal efficiency (55%) was obtained for the PP-T membrane in a comparison with the PP-N membranes without the talc additive (efficiency 43%). It can be concluded that for tested membranes, the magnitude of obtained permeate flux mainly determines the thermal efficiency of the MD process.

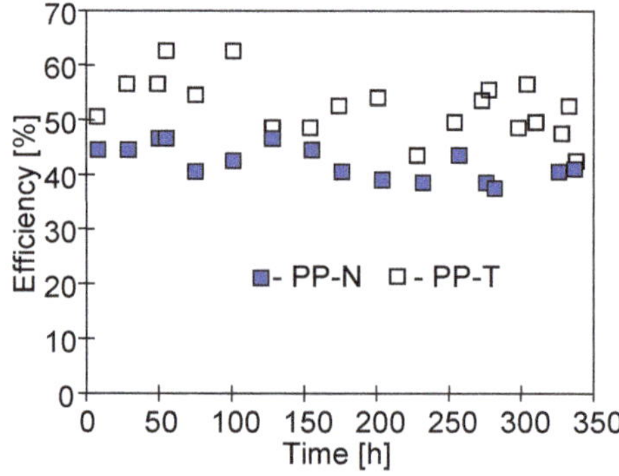

Figure 8. The changes of thermal efficiency of used membrane modules during the MD process.

3.3. Membrane Performance

A polymer with the linear structure formed on the surface of PP-N membranes can reduce the tensile strength of the membrane matrix. As a result, multiple cracks, probably formed during the membrane production, were observed on the external surface (Figure 2a). A long-term annealing of PP enables the rearrangement of PP chains in the membrane matrix, which improves the lamellae orientation and uniformity [61]; hence, further damages/cracks can be expected during the MD process.

The performed SEM examinations of membrane samples collected from MD modules after 350 h of their exploitation confirmed that a considerable degradation took place in the membranes without the talc addition. The external surface of PP-N membrane with visible numerous cracks is shown in Figure 9a. A number and the dimensions of these cracks are definitely larger in comparison to those observed on the surface of new membranes (Figure 2a). An enlarged image of cracks (Figure 9b) shows that the linear structures of PP chains undergo, not only a breakage but also a split of the membrane surface along the formed structures. These kinds of damages on the lumen side of PP-N membrane were found to occur to a lesser degree (Figure 9c). However, on this side the distillate flows at a low temperature (below 313 K). This confirms an assumption that a high feed temperature is the main reason for a stress that resulted in damage of the surface structure in the tested PP-N membranes. A phenomenon of thermal extension generates the stress also inside the PP-T membrane. However, talc addition improved the cellular structure, which is more resistant on the stress than a linear structure, and thus, the degradation of the membrane matrix was not observed (Figure 9d). The SEM examinations of PP-N membrane cross-sections revealed that a lack of cracks inside the capillary wall was due to the same reason. A fact, that the degradation of PP N membrane structure was limited to its external surfaces was also confirmed by the results of the MD process, in particular, by low values of electric conductivity of the obtained distillate (Figure 4).

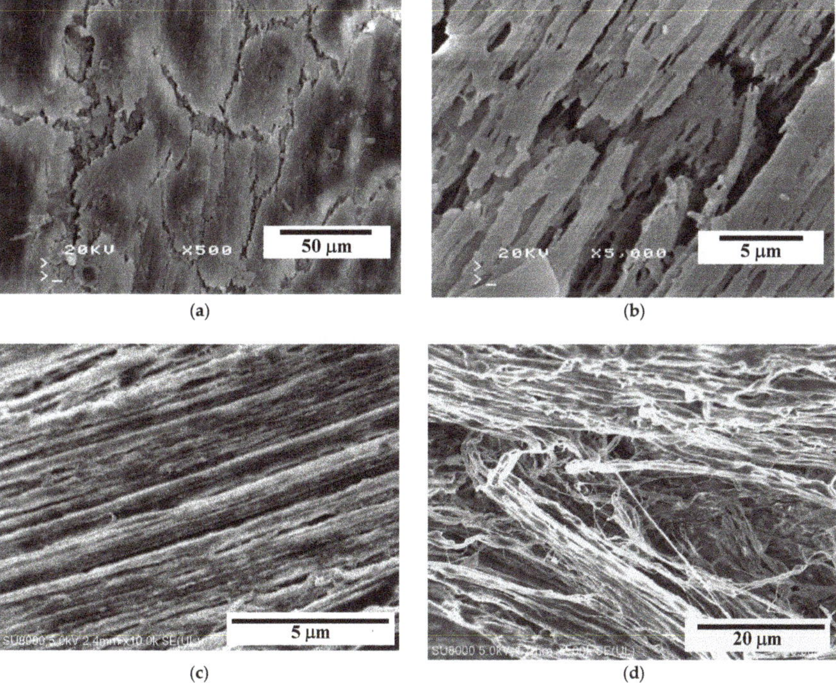

Figure 9. SEM images of PP membranes after MD process. Membrane PP-N: (**a**) external surface; (**b**) external surface magnification; (**c**) internal surface; (**d**) external surface of PP-T membrane.

Based on the above results, it can be concluded that a cellular structure occurring inside the capillary wall is definitely more resistant to the stress formed in the membrane in comparison with a linear structure of polymer formed on the membrane surfaces. This conclusion was confirmed by the results of tensile tests (Table 2). The values of all the studied parameters (Young's modulus, elongation at break and tensile strength) were better for membranes blended with talc, having a mainly sponge-like structure in comparison to that observed for PP-N membranes.

Table 2. The mechanical properties of studied PP membranes.

Membrane	Young's Modulus [MPa]	Elongation at Break [%]	Tensile Strength [MPa]
PP-N	106.1 +/− 18.3	154.5 +/− 26.8	1.98 +/− 0.13
PP-T	135.5 +/− 12.8	172.2 +/− 19.7	2.55 +/− 0.14

The mechanical strength of MD membranes is as important as their resistance to wetting. Even small cracks can cause a significant leakage of the feed. In the case of PP membranes, their strength depends, to a large extent on the degree of crystallinity and the structures that constitute the polymer chains.

The addition of talc accelerates the formation of crystallites and decreases the spherulite dimensions, thus improves the thermal and mechanical properties of polymer matrix [39,49]. The results of DSC analysis (Table 3) indicated the increase of melting enthalpy (crystallinity-X) and the crystallization temperature, which confirmed that talc is a good nucleating agent. However, the degradation processes occur during MD module exploitation, which would result in a remarkable deterioration in the membranes properties. The high feed temperature applied in the MD process can accelerate the thermal degradation of polypropylene [36,50]. The degree of degradation of the PP membrane matrix was determined in the DSC studies. In each case, it was observed that the membranes after the MD process exhibited a reduction of T_m values, which was larger (from 164.6 to 157.2 °C) in the case of PP-N membranes (Table 3). It has been reported that the thermodynamic melting temperature of semicrystalline polymers decreases as the molecular weight decreases and/or as the number of defects increases with an increase in branching and/or crosslinking [51].

Table 3. The results of differential scanning calorimetry (DSC) analysis of new membranes and membrane samples after MD studies.

Membrane	T_m [°C]	ΔH_m [J/g]	T_C [°C]	ΔH_C [J/g]	X [%]
PP-N	164.6	91.2	116.1	82.3	44
PP-N (MD)	157.2	84.9	117.1	68.4	41
PP-T	163.7	117.1	119.1	93.2	56
PP-T (MD)	162.7	109.8	114.6	91.7	53

The differences in composition and crystallinity can be also determined by XRD studies. For the membranes studied, five peaks (14.1, 16.9, 18.5, 21.4 and 21.8°) characteristic of isotactic PP (iPP) were obtained [46]. In the case of PP-T membranes in the XRD spectrum (Figure 10b) there were additional peaks resulting from the presence of talc. These membranes obtained narrower peaks, which confirms that the addition of talc increases the degree of crystallinity of PP membranes (Table 4). Long-term heating of membranes in the MD process promotes recrystallization [62], which is confirmed by changes in full-width at half maximum (FWHM) values determined for membranes after the MD process.

Table 4. The values of full-width at half maximum (FWHM) of the peaks obtained during XRD analysis.

Membrane	Peak 14.1 [°]	Peak 16.9 [°]
PP-N	0.89 +/− 0.01	1.0 +/− 0.02
PP-N (MD)	0.72 +/− 0.01	0.74 +/− 0.02
PP-T	0.67 +/− 0.01	0.71 +/− 0.02
PP-T (MD)	0.69 +/− 0.01	0.68 +/− 0.01

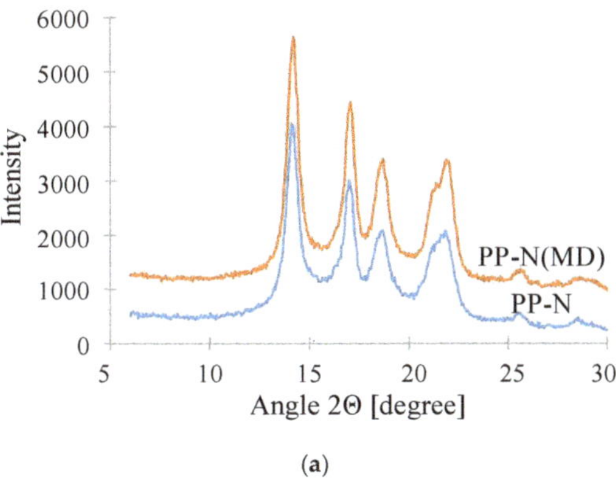

(**a**)

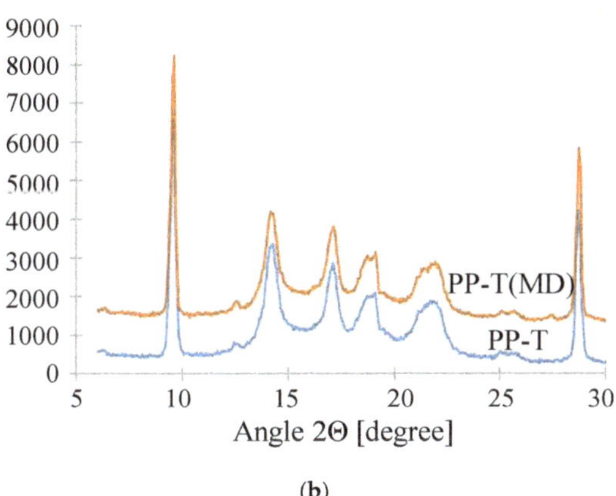

(**b**)

Figure 10. The results of XRD analysis. (**a**) membrane PP-N, (**b**) membrane PP-T.

3.4. Stability of Membrane Matrix

The membranes for membrane distillation are dry, i.e., the solutions separated by them cannot rinse the components from the membrane matrix. However, in the case of PP membranes after 50–100 h, the pores on the surface of the membranes become wet (partial wetting [50,59]), which allows an elution of components from the membrane matrix. For this reason it is very important that the fillers are permanently embedded in the membranes. The degree of talc leaching from membranes after 350 h of the MD process was evaluated by studying the changes in the concentration of the components forming the membrane matrix. The results obtained for elemental analysis are shown in Table 5.

Table 5. The results of elemental analysis.

Membrane	C [%]	O [%]	H [%]	O/C [%]
PP-N	85.613 +/− 0.811	0.976 +/− 0.009	14.423 +/− 0.126	1.14
PP-N (MD)	82.668 +/− 0.116	0.804 +/− 0.116	13.961 +/− 0.116	0.97
PP-T	79.788 +/− 0.116	3.783 +/− 0.281	13.312 +/− 0.015	4.74
PP-T (MD)	76.333 +/− 0.116	3.224 +/− 0.116	12.856 +/− 0.116	4.22

As can be expected, the membranes blended with talc exhibited higher oxygen content and less carbon. After the MD process, the concentration of all tested components was reduced; hence, the value of the oxygen to carbon ratio was calculated for the assessment of changes. The O/C value for the PP-T membrane decreased from 4.74% to 4.22%, which could indicate a slight loss of talc. However, a similar reduction in the O/C values was also obtained for the PP-N membrane containing no talc. For this reason, on the basis of these data it is difficult to conclude the leaching of talc. In the conducted research, the entire membrane wall was subjected to analysis, including also those areas that were not wetted. This makes it difficult to assess the changes in a thin outer layer of the wall. The sizes of most pore cells were in the range of 1–3 μm; thus, a similar thickness of the rinsed layer can be assumed. At a similar depth, the composition of the membrane wall is analyzed in the SEM-EDS and ATR-FTIR methods. An even thinner layer results in XPS testing.

The XPS studies analyze the composition of only a few external atomic layers; hence, the results may also include the composition of impurities, in our case Si, which occurs in various materials and could be washed out from the MD installation. However, in the conducted studies, the presence of Si was found only in the case of PP-T membranes (Figure 11). The content of elements detected in the tested membrane samples is shown in Table 6.

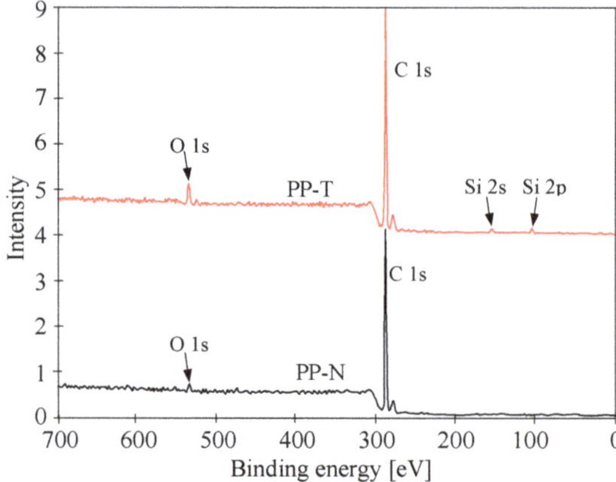

Figure 11. Results of XPS analysis surface of membrane samples (feed side) after 350 h of MD process duration.

Table 6. The results of EDS and XPS analysis of membrane samples.

Membrane	C [%]	O [%]	Si [%]	Mg [%]
PP-N (EDS)	98.3	0.9	-	-
PP-N (XPS)	98.1	1.9	-	-
PP-T (EDS)	93.1	2.7	2.2	1.8
PP-T (XPS)	91.8	4.2	2.5	1.4

The XPS test results can be disturbed by even small amounts of talc that could have deposited on the surface of the membranes during their production. Therefore, the results of tests carried out using the SEM-EDS method for the new membranes are provided (Table 6). It can be concluded that these results are similar to those obtained with the XPS method for membranes after 350 h of the MD process. This confirms that the leaching of talc from the PP-T membrane matrix is practically negligible. Moreover, it can be seen that PP-N membranes did not contain talc, although oxygen was detected in them. The presence of oxygen results from the degradation processes of PP, which results in the formation of e.g., aldehydes and ketones [63].

The changes in the membrane surface composition can be tested by the FTIR method. The results obtained for the new membranes PP-N and PP-T are shown in Figure 12. The obtained FTIR spectra have a methyl absorption band at 1375 cm^{-1} and 1450 cm^{-1} characteristic for PP and few peaks (809, 841, 899, 972, 997.1167 and 1219 cm^{-1}) attributed to the presence of iPP [64,65]. In the case of PP-N membranes, the significantly larger peaks had an intensity that was observed in the range of 1600–1800 cm^{-1}, which is characteristic for the carbonyl groups formed as a result of thermal oxidation [62]. Due to the high temperature of the TIPS process, the polymer degradation already occurs at the membrane production stage [66]. The addition of talc increases the thermal resistance of membranes; hence the degree of degradation of PP-T membranes was reduced. This is also a reason why the PP-T membranes have a larger contact angle (Figure 7).

The FTIR tests also allows for confirming the presence of talc in the membrane matrix, which show peaks observed in the range of 400–1200 cm^{-1} for the tested membranes (Figure 13). The appearance of the absorbance peak at 1018 cm^{-1} is attributed to the stretching vibrations of Si-O-Si [67]. Large peaks also occur at 660 cm^{-1} and wide in the 400–550 cm^{-1} ranges, which were not observed for the PP-N membrane. A comparison of the data in Figure 13, allows for us to conclude that the intensity of talc peaks is similar for PP-T and PP-T (MD) membranes, confirming the previous results indicating that talc did not rinse during long-term MD process studies.

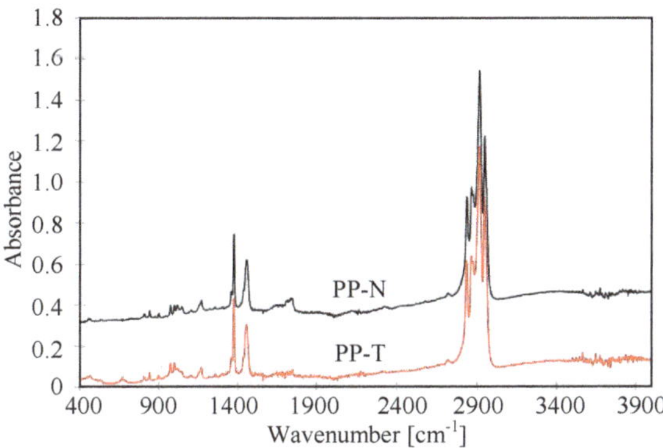

Figure 12. Results of FTIR analysis of tested PP-N and PP-T membranes.

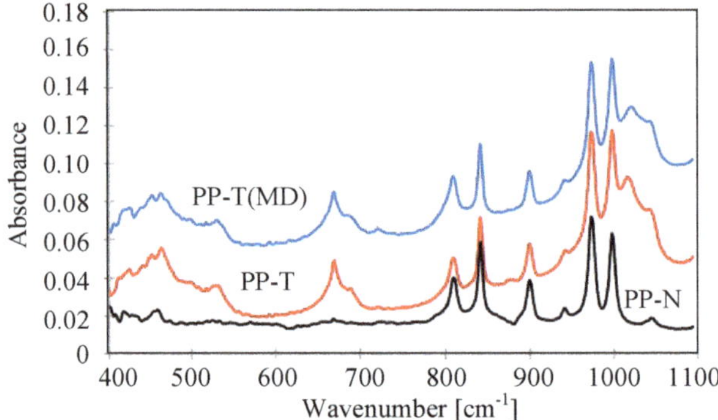

Figure 13. The results of FTIR analysis—peaks characteristic for talc. PP-T(MD) membrane sample colected from MD module after 350 h of the MD process.

4. Conclusions

The obtained MD results demonstrated that both types of PP membranes used in the work possess the appropriate properties for the MD process, such as a resistance to wetting and almost 100% salt retention. However, the additional tests of the membrane matrix stability revealed that only the PP membranes that were reinforced by talc addition had the properties suitable for application to pilot plant MD studies. Moreover, the MD tests carried out for 350 h did not allow for revealing the occurrence of membrane defects and for this purpose, an even longer MD study should be performed, when the industrial application of membranes is considered.

During the membrane formation via the TIPS method, the flow of melted polypropylene through a spinning nozzle caused a linear arrangement of the polymer chains. A cooling-down of the formed capillary in the coagulation bath proceeded rapidly, favoring the freezing of polymer linear structure. A result is a deterioration of the tensile strength manifested by a longitudinal cracking of polymer. The addition of talc (crystals nucleus) into a dope solution caused the disturbance in the linear arrangement of the polymer chains, and PP membranes with a surface morphology more akin to a sponge-like structure was formed. In this case, the MD process efficiency was higher by more than 10% in comparison to the utilization of membranes without talc addition.

In the MD process, the membranes are in contact with the feed at a high temperature, e.g., 353 K, which causes the formation of thermal stress resulting in the numerous cracks which were observed on the surface of membranes having a linear structure. Such problems were not found in the case of membranes where the linear arrangement of polymer chains was disturbed by the addition of talc into a dope solution.

It has been confirmed that the uniform dispersion of talc was achieved within the membrane formed by the TIPS method. Talc was permanently incorporated into the polypropylene matrix and did not leach from the PP membranes during the MD process.

Funding: This research was funded by National Science Centre, Poland, grant number 2018/29/B/ST8/00942.

Conflicts of Interest: The author declares no conflict of interest. The funders had no role in the design of the study; in the collection, analyses, or interpretation of data; in the writing of the manuscript, or in the decision to publish the results.

References

1. Bodell, B.R. Silicone Rubber Vapor Diffusion in Saline Water Desalination. U.S. Patent 285032, 3 June 1963.
2. Lizeth, D.; Mendez, M.; Castel, C.; Lemaitre, C.; Favre, E. Membrane distillation (MD) processes for water desalination applications. Can dense selfstanding membranes compete with microporous hydrophobic materials? *Chem. Eng. Sci.* **2018**, *188*, 84–96.

3. Winter, D.; Koschikowski, J.; Wieghaus, M. Desalination using membrane distillation: Experimental studies on full scale spiral wound modules. *J. Membr. Sci.* **2011**, *375*, 104–112. [CrossRef]
4. Guillén-Burrieza, E.; Blanco, J.; Zaragoza, G.; Alarcón, D.-C.; Palenzuela, P.; Ibarra, M.; Gernjak, W. Experimental analysis of an air gap membrane distillation solar desalination pilot system. *J. Membr. Sci.* **2011**, *379*, 386–396. [CrossRef]
5. Duong, H.C.; Chivas, A.R.; Nelemans, B.; Duke, M.; Gray, S.; Cath, T.Y.; Nghiem, L.D. Treatment of RO brine from CSG produced water by spiral-wound air gap membrane distillation—A pilot study. *Desalination* **2015**, *366*, 121–129. [CrossRef]
6. Schwantes, R.; Cipollina, A.; Gross, F.; Koschikowski, J.; Pfeifle, D.; Rolletschek, M.; Subiela, V. Membrane distillation: Solar and waste heat driven demonstration plants for desalination. *Desalination* **2013**, *323*, 93–106. [CrossRef]
7. Qtaishat, M.R.; Banat, F. Desalination by solar powered membrane distillation systems. *Desalination* **2013**, *308*, 186–197. [CrossRef]
8. Eykens, L.; De Sitter, K.; Dotremont, C.; Pinoy, L.; Van der Bruggen, B. Membrane synthesis for membrane distillation: A review. *Sep. Purif. Technol.* **2017**, *182*, 36–51. [CrossRef]
9. Rezaeia, M.; Warsinger, D.M.; Lienhard V, J.H.; Samhaber, W.M. Wetting prevention in membrane distillation through superhydrophobicity and recharging an air layer on the membrane surface. *J. Membr. Sci.* **2017**, *530*, 42–52. [CrossRef]
10. Gryta, M. The application of polypropylene membranes for production of fresh water from brines by membrane distillation. *Chem. Pap.* **2017**, *71*, 775–784. [CrossRef]
11. Adnan, S.; Hoang, M.; Wang, H.T.; Xie, Z.L. Commercial PTFE membranes for membrane distillation application: Effect of microstructure and support material. *Desalination* **2012**, *284*, 297–308. [CrossRef]
12. Eykens, L.; De Sitter, K.; Dotremont, C.; Pinoy, L.; Van der Bruggen, B. Characterization and performance evaluation of commercially available hydrophobic membranes for direct contact membrane distillation. *Desalination* **2016**, *392*, 63–73. [CrossRef]
13. Wang, P.; Chung, T.S. Recent advances in membrane distillation processes: Membrane development, configuration design and application exploring. *J. Membr. Sci.* **2015**, *474*, 39–56. [CrossRef]
14. González, D.; Amigo, J.; Suárez, F. Membrane distillation: Perspectives for sustainable and improved desalination. *Renew. Sustain. Energy Rev.* **2017**, *80*, 238–259. [CrossRef]
15. Khalifa, A.; Ahmad, H.; Antar, M.; Laoui, T.; Khayet, M. Experimental and theoretical investigations on water desalination using direct contact membrane distillation. *Desalination* **2017**, *404*, 22–34. [CrossRef]
16. Rezaei, M.; Warsinger, D.M.; Lienhard V, J.H.; Duke, M.C.; Matsuura, T.; Samhaber, W.M. Wetting phenomena in membrane distillation: Mechanisms, reversal, and prevention. *Water Res.* **2018**, *139*, 329–352. [CrossRef]
17. Chamani, H.; Matsuura, T.; Rana, D.; Lan, C.Q. Modeling of pore wetting in vacuum membrane distillation. *J. Membr. Sci.* **2019**, *572*, 332–342. [CrossRef]
18. Thomas, N.; Mavukkandy, M.O.; Loutatidou, S.; Arafat, H.A. Membrane distillation research & implementation: Lessons from the past five decades. *Sep. Purif. Technol.* **2017**, *189*, 108–127.
19. Lee, E.-J.; An, A.K.; He, T.; Woo, Y.C.; Shon, H.K. Electrospun nanofiber membranes incorporating fluorosilane-coated TiO_2 nanocomposite for direct contact membrane distillation. *J. Membr. Sci.* **2016**, *520*, 145–154. [CrossRef]
20. Ali, A.; Tsai, J.-H.; Tung, K.-L.; Drioli, E.; Macedonio, F. Designing and optimization of continuous direct contact membrane distillation process. *Desalination* **2018**, *426*, 97–107. [CrossRef]
21. Eykens, L.; Hitsov, I.; DeSitter, K.; Dotremont, C.; Pinoy, L.; Nopens, I.; Van der Bruggen, B. Influence of membrane thickness and process conditions on direct contact membrane distillation at different salinities. *J. Membr. Sci.* **2016**, *498*, 353–364. [CrossRef]
22. Guillen-Burrieza, E.; Ruiz-Aguirre, A.; Zaragoza, G.; Arafat, H.A. Membrane fouling and cleaning in long term plant-scale membrane distillation operations. *J. Membr. Sci.* **2014**, *68*, 360–372. [CrossRef]
23. Himma, N.F.; Anisah, S.; Prasetya, N.; Wenten, I.G. Advances in preparation, modification, and application of polypropylene membrane. *Polym. Eng.* **2016**, *36*, 329–362. [CrossRef]
24. Eykens, L.; De Sitter, K.; Dotremont, C.; Pinoy, L.; Van der Bruggen, B. Coating techniques for membrane distillation: An experimental assessment. *Sep. Purif. Technol.* **2018**, *193*, 38–48. [CrossRef]
25. Cui, Z.; Zhang, Y.; Li, X.; Wang, X.; Drioli, E.; Wang, Z.; Zhao, S. Optimization of novel composite membranes for water and mineral recovery by vacuum membrane distillation. *Desalination* **2018**, *440*, 39–47. [CrossRef]

26. Ali, A.; Criscuoli, A.; Macedonio, F.; Drioli, E. A comparative analysis of flat sheet and capillary membranes for membrane distillation applications. *Desalination* **2019**, *456*, 1–12. [CrossRef]
27. Silva, T.L.S.; Morales-Torres, S.; Figueiredo, J.L.; Silva, A.M.T. Multi-walled carbon nanotube/PVDF blended membranes with sponge- and finger-like pores for direct contact membrane distillation. *Desalination* **2015**, *357*, 233–245. [CrossRef]
28. Fahmey, M.S.; El-Aassar, A.-H.M.; Abo-Elfadel, M.M.; Orabi, A.S.; Das, R. Comparative performance evaluations of nanomaterials mixed polysulfone: A scale-up approach through vacuum enhanced direct contact membrane distillation for water desalination. *Desalination* **2019**, *451*, 111–116. [CrossRef]
29. Baghbanzadeh, M.; Rana, D.; Lan, C.Q.; Matsuura, T. Effects of hydrophilic silica nanoparticles and backing material in improving the structure and performance of VMD PVDF membranes. *Sep. Purif. Technol.* **2016**, *157*, 60–71. [CrossRef]
30. Makhlouf, A.; Satha, H.; Frihi, D.; Gherib, S.; Seguela, R. Optimization of the crystallinity of polypropylene/submicronic-talc composites: The role of filler ratio and cooling rate. *eXPRESS Polym. Lett.* **2016**, *10*, 237–247. [CrossRef]
31. Efome, J.E.; Baghbanzadeh, M.; Rana, D.; Matsuura, T.; Lan, C.Q. Effects of superhydrophobic SiO_2 nanoparticles on the performance of PVDF flat sheet membranes for vacuum membrane distillation. *Desalination* **2015**, *373*, 47–57. [CrossRef]
32. Santoro, S.; Vidorreta, I.; Coelhoso, I.; Lima, J.C.; Desiderio, G.; Lombardo, G.; Drioli, E.; Mallada, R.; Crespo, J.; Criscuoli, A.; et al. Experimental Evaluation of the Thermal Polarization in Direct Contact Membrane Distillation Using Electrospun Nanofiber Membranes Doped with Molecular Probes. *Molecules* **2019**, *24*, 638. [CrossRef]
33. Kujawa, J.; Al-Gharabli, S.; Kujawski, W.; Knozowska, K. Molecular Grafting of Fluorinated and Nonfluorinated Alkylsiloxanes on Various Ceramic Membrane Surfaces for the Removal of Volatile Organic Compounds Applying Vacuum Membrane Distillation. *ACS Appl. Mater. Interfaces* **2017**, *9*, 6571–6590. [CrossRef]
34. Hubadillah, S.K.; Tai, Z.S.; Othman, M.H.D.; Harun, Z.; Jamalludin, M.R.; Rahmana, M.A.; Jaafar, J.; Ismail, A.F. Hydrophobic ceramic membrane for membrane distillation: A mini review on preparation, characterization, and applications. *Sep. Purif. Technol.* **2019**, *217*, 71–84. [CrossRef]
35. Ashoor, B.B.; Mansour, S.; Giwa, A.; Dufour, V.; Hasan, S.W. Principles and applications of direct contact membrane distillation (DCMD): A comprehensive review. *Desalination* **2016**, *398*, 222–246. [CrossRef]
36. Gryta, M.; Grzechulska-Damszel, J.; Markowska, A.; Karakulski, K. The influence of polypropylene degradation on the membrane wettability during membrane distillation. *J. Membr. Sci.* **2009**, *326*, 493–502. [CrossRef]
37. Wang, Y.-J.; Zhao, Z.-P.; Xi, Z.-Y.; Yan, S.-Y. Microporous polypropylene membrane prepared via TIPS using environment-friendly binary diluents and its VMD performance. *J. Membr. Sci.* **2018**, *548*, 332–344. [CrossRef]
38. Tang, N.; Han, H.; Yuan, L.; Zhang, L.; Wang, X.; Cheng, P. Preparation of a hydrophobically enhanced antifouling isotactic polypropylene/silicone dioxide flat-sheet membrane via thermally induced phase separation for vacuum membrane distillation. *J. Appl. Polym. Sci.* **2015**, *132*, 1–11. [CrossRef]
39. Xu, K.; Cai, Y.; Hassankiadeh, N.T.; Cheng, Y.; Li, X.; Wang, X.; Wang, Z.; Drioli, E.; Cui, Z. ECTFE membrane fabrication via TIPS method using ATBC diluent for vacuum membrane distillation. *Desalination* **2019**, *456*, 13–22. [CrossRef]
40. Zhang, X.; Zhang, D.; Liu, T. Influence of Nucleating Agent on Properties of Isotactic Polypropylene. *Energy Procedia* **2012**, *17*, 1829–1835. [CrossRef]
41. Tang, N.; Jia, Q.; Zhang, H.; Li, J.; Cao, S. Preparation and morphological characterization of narrow pore size distributed polypropylene hydrophobic membranes for vacuum membrane distillation via thermally induced phase separation. *Desalination* **2010**, *256*, 27–36. [CrossRef]
42. Lin, Y.K.; Chen, G.; Yang, J.; Wang, X.L. Formation of isotactic polypropylene membranes with bicontinuous structure and good strength via thermally induced phase separation method. *Desalination* **2009**, *236*, 8–15. [CrossRef]
43. Tang, N.; Feng, C.; Han, H.; Hua, X.; Zhang, L.; Xiang, J.; Cheng, P.; Du, W.; Wang, X. High permeation flux polypropylene/ethylene vinyl acetate co-blending membranes via thermally induced phase separation for vacuum membrane distillation desalination. *Desalination* **2016**, *394*, 44–55. [CrossRef]

44. Zhao, J.; Shi, L.; Loh, C.H.; Wang, R. Preparation of PVDF/PTFE hollow fiber membranes for direct contact membrane distillation via thermally induced phase separation method. *Desalination* **2018**, *430*, 86–97. [CrossRef]
45. Matsuyama, M.; Okafuji, H.; Maki, T.; Teramoto, M.; Kubota, N. Preparation of polyethylene hollow fiber membrane via thermally induced phase separation. *J. Membr. Sci.* **2003**, *223*, 119–126. [CrossRef]
46. Li, M.; Qi, Y.; Zhao, Z.; Xiang, Z.; Liao, X.; Niu, Y.; Kong, M. Morphology evolution and crystalline structure of controlled-rheology polypropylene in micro-injection molding. *Polym. Adv. Technol.* **2016**, *27*, 494–503. [CrossRef]
47. Ammar, O.; Bouaziz, Y.; Haddar, N.; Mnif, N. Talc as Reinforcing Filler in Polypropylene Compounds: Effect on Morphology and Mechanical Properties. *Polym Sci.* **2017**, *3*, 1–7.
48. Alvarez, V.A.; Pérez, C.J. Effect of different inorganic filler over isothermal and non-isothermal crystallization of polypropylene homopolymer. *J. Therm. Anal. Calorim.* **2012**, *107*, 633–643. [CrossRef]
49. Luo, B.; Zhang, J.; Wang, X.; Zhou, Y.; Wen, J. Effects of nucleating agents and extractants on the structure of polypropylene microporous membranes via thermally induced phase separation. *Desalination* **2006**, *192*, 142–150. [CrossRef]
50. Gryta, M. Long-term performance of membrane distillation process. *J. Membr. Sci.* **2005**, *265*, 153–159. [CrossRef]
51. Ursino, C.; Di Nicolò, E.; Gabriele, B.; Criscuoli, A.; Figoli, A. Development of a novel perfluoropolyether (PFPE) hydrophobic/hydrophilic coated membranes for water treatment. *J. Membr. Sci.* **2019**, *581*, 58–71. [CrossRef]
52. Kalčíková, G.; Marolt, G.; Kokalj, A.J.; Gotvajn, A.Ž. The use of multiwell culture plates in the duckweed toxicity test—A case study on Zn nanoparticles. *New Biotechnol.* **2018**, *47*, 67–72. [CrossRef]
53. Rekulapally, R.; Murthy Chavali, L.N.; Idris, M.M.; Singh, S. Toxicity of TiO_2, SiO_2, ZnO, CuO, Au and Ag engineered nanoparticles on hatching and early nauplii of Artemia sp. *PeerJ* **2019**, *6*, e6138. [CrossRef]
54. Boffetta, P.; Mundt, K.A.; Thompson, W.J. The epidemiologic evidence for elongate mineral particle (EMP)-related human cancer risk. *Toxicol. Appl. Pharmacol.* **2018**, *361*, 100–106. [CrossRef]
55. Drechsel, D.A.; Barlow, C.A.; Bare, J.L.; Jacobs, N.F.; Henshaw, J.L. Historical evolution of regulatory standards for occupational and consumer exposures to industrial talc. *Regul. Toxicol. Pharmacol.* **2018**, *92*, 251–267. [CrossRef]
56. Chang, C.-J.; Yang, Y.-H.; Chen, P.-C.; Peng, H.-Y.; Lu, Y.-C.; Song, S.-R.; Yang, H.-Y. Stomach Cancer and Exposure to Talc Powder without Asbestos via Chinese Herbal Medicine: A Population-Based Cohort Study. *Int. J. Environ. Res. Public Health* **2019**, *16*, 717. [CrossRef] [PubMed]
57. Moulik, S.; Kumar, F.D.; Archana, K.; Sridhar, S. Enrichment of hydrazine from aqueous solutions by vacuum membrane distillation through microporous polystyrene membranes of enhanced hydrophobicity. *Sep. Purif. Technol.* **2018**, *203*, 159–167. [CrossRef]
58. Li, L.; Li, W.; Geng, L.; Chen, B.; Mi, H.; Hong, K.; Peng, X.; Kuang, T. Formation of stretched fibrils and nanohybrid shish-kebabs in isotactic polypropylene-based nanocomposites by application of a dynamic oscillatory shear. *Chem. Eng. J.* **2018**, *348*, 546–556. [CrossRef]
59. Gryta, M. Influence of polypropylene membrane surface porosity on the performance of membrane distillation process. *J. Membr. Sci.* **2007**, *287*, 67–78. [CrossRef]
60. Li, Z.; Rana, D.; Wang, Z.; Matsuura, T.; Lan, C.Q. Synergic effects of hydrophilic and hydrophobic nanoparticles on performance of nanocomposite distillation membranes: An experimental and numerical study. *Sep. Purif. Technol.* **2018**, *202*, 45–58. [CrossRef]
61. Sadeghi, F.; Ajji, A.; Carreau, P.J. Analysis of microporous membranes obtained from polypropylene films by stretching. *J. Membr. Sci.* **2007**, *292*, 62–71. [CrossRef]
62. Lv, Y.; Huang, Y.; Kong, M.; Li, G. Improved thermal oxidation stability of polypropylene films in the presence of β-nucleating agent. *Polym. Test.* **2013**, *32*, 179–186. [CrossRef]
63. Fontanella, S.; Bonhomme, S.; Brusson, J.-M.; Pitteri, S.; Samuel, G.; Pichon, G.; Lacoste, J.; Fromageot, D.; Lemaire, J.; Delort, A.-M. Comparison of biodegradability of various polypropylene films containing pro-oxidant additives based on Mn, Mn/Fe or Co. *Polym. Degrad. Stabil.* **2013**, *98*, 875–884. [CrossRef]
64. Echeverrigaray, S.G.; Cruz, R.C.D.; Oliveira, R.V.B. Reactive processing of a non-additivated isotactic polypropylene: Mechanical and morphological properties on molten and solid states. *Polym. Bull.* **2013**, *70*, 1237–1250. [CrossRef]

65. He, P.; Xiao, Y.; Zhang, P.; Xing, C.; Zhu, N.; Zhu, X.; Yan, D. Thermal degradation of syndiotactic polypropylene and the influence of stereoregularity on the thermal degradation behaviour by in situ FTIR spectroscopy. *Polym. Degrad. Stab.* **2005**, *88*, 473–479. [CrossRef]
66. Qian, S.; Igarashi, T.; Nitta, K. Thermal degradation behavior of polypropylene in the melt state: Molecular weight distribution changes and chain scission mechanism. *Polym. Bull.* **2011**, *67*, 1661–1670. [CrossRef]
67. Zhang, H.; Li, B.; Sun, D.; Miao, X.; Gu, Y. SiO_2-PDMS-PVDF hollow fiber membrane with high flux for vacuum membrane distillation. *Desalination* **2018**, *429*, 33–43. [CrossRef]

Article

Porous Hydrophobic–Hydrophilic Composite Hollow Fiber and Flat Membranes Prepared by Plasma Polymerization for Direct Contact Membrane Distillation

Ashok K. Sharma [1], Adam Juelfs [1], Connor Colling [1], Saket Sharma [1], Stephen P. Conover [1,*], Aishwarya A. Puranik [2], John Chau [2], Lydia Rodrigues [2] and Kamalesh K. Sirkar [2,*]

1 Applied Membrane Technology Inc., 11558 Encore Circle, Minnetonka, MN 55343, USA; aksharma@appliedmembranetech.com (A.K.S.); juelf002@umn.edu (A.J.); ccolling@appliedmembranetech.com (C.C.); saketsharma@appliedmembranetech.com (S.S.)
2 Otto York Department of Chemical and Materials Engineering, New Jersey Institute of Technology, University Heights, Newark, NJ 07102, USA; aap253@njit.edu (A.A.P.); jc56@njit.edu (J.C.); lr59@njit.edu (L.R.)
* Correspondence: spconover@appliedmembranetech.com (S.P.C.); sirkar@njit.edu (K.K.S.); Tel.: +1-952-933-5121 (S.P.C.); +1-973-596-8447 (K.K.S.); Fax: +1-952-933-8839 (S.P.C.); +1-973-642-4854 (K.K.S.)

Abstract: High water vapor flux at low brine temperatures without surface fouling is needed in membrane distillation-based desalination. Brine crossflow over surface-modified hydrophobic hollow fiber membranes (HFMs) yielded fouling-free operation with supersaturated solutions of scaling salts and their precipitates. Surface modification involved an ultrathin porous polyfluorosiloxane or polysiloxane coating deposited on the outside of porous polypropylene (PP) HFMs by plasma polymerization. The outside of hydrophilic MicroPES HFMs of polyethersulfone was also coated by an ultrathin coating of porous plasma-polymerized polyfluorosiloxane or polysiloxane rendering the surface hydrophobic. Direct contact membrane distillation-based desalination performances of these HFMs were determined and compared with porous PP based HFMs. Salt concentrations of 1, 10, and 20 wt% were used. Leak rates were determined at low pressures. Surface and cross-sections of two kinds of coated HFMs were investigated by scanning electron microscopy. The HFMs based on water-wetted MicroPES substrate offered a very thin gas gap in the hydrophobic surface coating yielding a high flux of 26.4–27.6 kg/m^2-h with 1 wt% feed brine at 70 °C. The fluxes of HFMs on porous PP substrates having a long vapor diffusion path were significantly lower. Coated HFM performances have been compared with flat hydrophilic membranes of polyvinylidene fluoride having a similar plasma-polymerized hydrophobic polyfluorosiloxane coating.

Keywords: membrane distillation; composite membrane; plasma-polymerized hydrophobic fluorosiloxane coating; hydrophilic porous hollow-fiber substrate

Citation: Sharma, A.K.; Juelfs, A.; Colling, C.; Sharma, S.; Conover, S.P.; Puranik, A.A.; Chau, J.; Rodrigues, L.; Sirkar, K.K. Porous Hydrophobic–Hydrophilic Composite Hollow Fiber and Flat Membranes Prepared by Plasma Polymerization for Direct Contact Membrane Distillation. *Membranes* **2021**, *11*, 120. https://doi.org/10.3390/membranes11020120

Academic Editor: Alessandra Criscuoli

Received: 31 December 2020
Accepted: 4 February 2021
Published: 8 February 2021

1. Introduction

Water evaporation through a porous hydrophobic membrane is being intensively investigated for water desalination and a few other water treatment processes. This process, termed membrane distillation (MD), has several variants depending on how water evaporated from hot brine on the feed side of the membrane is condensed on the other side of the membrane. We are here concerned primarily with the direct contact membrane distillation (DCMD) process, where the evaporated water is condensed by a flowing stream of cold distilled water on the other side of the membrane. There are excellent and comprehensive review articles on various aspects of MD and DCMD [1–7], the last one being the earliest containing a mini-review of MD. There may be occasional literature references here to other MD varieties: vacuum membrane distillation (VMD), air gap membrane distillation (AGMD), and sweep gas membrane distillation (SGMD).

In DCMD, a porous gas-filled hydrophobic membrane exists between hot brine and cold distillate streams. Water evaporated at the hot brine-membrane interface is transported through gas-filled pores to the other side where it condenses in the flowing cold distillate. Water vapor flux depends on the physical properties of the porous membrane, the temperatures of the two water streams, extent of temperature polarization in the two flowing liquids, the rate of conductive heat loss as well as any membrane surface fouling. Additional issues of importance involve the extent of water recovery. A few major recent developments in these and other aspects of MD will be identified first before we converge on the subject of this work involving water vapor flux in surface-modified hollow fiber membranes.

Water flux in MD depends on the water vapor pressures and therefore the temperatures of water in contact with the two sides of the membrane. Conventionally, in MD, the hydrophobic membrane is contacted with hot brine from an external source to generate water vapor at the water-membrane interface. Such a process does lead to temperature polarization wherein the bulk brine temperature is higher than the temperature at the brine-membrane interface, which leads to a lower water vapor pressure-based driving force. To enhance MD flux, Chen et al. [8] pursued breaking hydrogen bonds in water via gold nanoparticles-adsorbed ceramic rods (AuNPs@CRs) and enhancing the water vapor pressure. Techniques that use other sources to heat the membrane-water interface region e.g., "self-heating" membrane by a high-frequency magnetic field [9], electrically conducting MD system heated by Ohmic resistance at alternating currents [10], avoid the need for preheating bulk feed brine and, therefore, claim lower intrinsic heat consumption. These techniques also do not suffer from negative consequences of temperature polarization since membrane surface temperature is higher than bulk brine temperature (on the other hand, there is heat loss to the bulk water). However, such claims are not relevant for cases where DCMD is employed to utilize waste heat already available in a hot brine (e.g., produced water [11]) or where the feed brine is heated via a solar collector. Enhancement of water vapor flux and stable operation then become more important.

A major problem in DCMD is membrane fouling, which can lead to reduction in salt rejection as well as reduced flux. Membrane pores may undergo wetting and, therefore, the process may need to be shut down and the membrane regenerated. Membrane fouling may be due to a variety of phenomena: scaling salt fouling, silica fouling, organic fouling, particle fouling, colloidal fouling, or biofouling. Particular sources of fouling may be mitigated under certain conditions as shown in the following studies.

To prevent fouling by scaling salt precipitates, the following technique using hollow fiber membranes (HFMs) appears to have been successful [12–14]. Hot feed brine was directed in cross flow across the outside surface of porous hydrophobic polypropylene (PP) HFMs in rectangular modules. The HFMs had a thin highly porous hydrophobic plasma-polymerized fluorosiloxane coating on the outer surface. In laboratory studies as well as pilot plant operation to concentrate seawater, the near-superhydrophobic membrane surface (contact angle ~140° [13]) was not fouled; although precipitates of scaling salts were floating all around, water flux was unaffected [12–14]. HFM-based cross flow modules specifically designed to have hot brine flow over and around HFMs create numerous flow separation points, which in turn generate secondary flows, which spontaneously scrub the membrane surface and prevent membrane fouling. In treatment of produced water containing a variety of salts as well as silica, 80% of water was recovered; no silica fouling of the membrane was observed [11].

Water-soluble agents, such as alcohols and surfactants, which reduce the surface tension of the solution, can easily wet the pores of most hydrophobic membranes in DCMD; surfactants take time to wet the pores but alcohols do it instantly [15]. Porous polyvinylidene fluoride (PVDF) membrane surfaces were modified by CF_4 plasma treatment to make superhydrophobic membrane surface [16], which exhibited negative charges on the surface, as well as stable MD performance in concentrating anionic surfactant-emulsified waste water; however, the membrane suffered from severe wetting when cationic surfactant was used. Stick water waste from meat rendering operations (rich in proteins, fats, and minerals) end up wetting hydrophobic membrane pores due to the fats. The fats got attached to membrane pores and facilitated liquid flow through the hydrophobic polytetrafluoroethylene (PTFE) membrane, which showed a total loss of flux within 0.5 h [17].

However, commercial hydrophobic PTFE membranes with a hydrophilic polyurethane (PU) surface layer were used successfully in DCMD on real poultry, fish, and bovine stick waters [17]; a metal microfiltration membrane was used to capture fats prior to DCMD. Tang et al. [18] showed how hydroxide ion generation, driven by water electrolysis on the electrically conducting membrane surface consisting of carbon nanotubes and a polyvinyl alcohol layer over a hydrophobic layer, can be used to efficiently dissolve silicate scaling that developed during the process of desalinating the geothermal brine by MD, negating the need for chemical cleaning. Note: a previous study employing coated HFMs with porous polyfluorosiloxane coatings operated with crossflow mode of hot produced water containing silica did not show any fouling [11].

It is known that in one pass of the brine through a DCMD unit, one cannot evaporate more than a limited fraction of the feed water. He et al. [19] pointed out that one can evaporate anywhere between 7 and 9% of the feed water in one pass, depending on the thermal efficiency of the DCMD unit for hot incoming brine at 90 °C and exiting at 35 °C. This has also been emphasized by Winter et al. [20]. Therefore He et al. [19] developed a concentration cascade of DCMD units with heat exchangers to achieve water recovery upwards of 60%. Such a cascade involves integration of multiple DCMD units with multiple heat exchangers to recycle the heat.

In AGMD, one usually achieves a higher thermal efficiency than DCMD due to the air gap, which also reduces the flux. By using an AGMD module having two adjoining and well-mixed sets of hollow fibers—one porous and hydrophobic for MD, and the other being a solid hollow fiber with a water-impervious wall to allow a coolant to flow through its fiber bore as it condenses the distillate vapor in the intervening shell-side space—a much higher module production rate can be achieved [21] than that in other AGMD designs. However, fractional water evaporation achieved in one pass is still limited. A flat plate-based new configuration called Feed Gap Air Gap MD (FGAGMD) has been designed [22] in which the feed brine is separated from the heating stream by a heat exchange surface; the cooling liquid is also separated from the permeate channel as in AGMD. Such a channel arrangement has allowed achievement of high one pass water recovery varying between 32–93% depending on the feed and its salt concentration.

All techniques in DCMD rely so far on a porous hydrophobic membrane layer to evaporate water on one surface and condense water at the other surface. The thinner this membrane layer is, the higher the flux [23]; a thinner membrane layer leads to a higher conductive flux loss, which depends however on the temperature difference between the two sides. With a very thin hydrophobic layer and a low ΔT between the two surfaces, conductive heat loss can be minimized, and the thermal efficiency can be high in DCMD (as much as 86%); yet the flux can still be reasonable [24]. Such a thin porous hydrophobic layer may be realized in a number of ways, e.g., plasma polymerization, polymer coating/grafting, and using hydrophobic fluorinated surface modifying macromolecules (SMMs) in the casting solvent used to fabricate a porous hydrophilic membrane [25–28].

A thin layer of polyvinyl alcohol (PVA) was coated successfully on a porous PTFE membrane by solution dipping followed by a cross-linking step to reduce the gas-filled pore length by a hydrophilic coating of polyvinyl alcohol. The water flux achieved with a 30 g/L salt solution was 12.2 L/m^2-h with a salt rejection of 99.9% for a feed temperature of 60 °C and distillate temperature of 17 °C [29]. A novel dual-layer flat membrane consisting of a thin hydrophobic top-layer of PVDF and a relatively thick hydrophilic PVDF-PVA sub-layer was fabricated using the non-solvent thermally induced phase separation (NTIPS) method [30]; the water vapor flux achieved was quite high.

A porous hydrophilic membrane prepared by casting from a NMP solution of polysulfone blended with polyvinyl pyrrolidone was converted into a hydrophobic membrane by treating it with CF_4 plasma [31]. The gaseous plasma modified not only the exposed membrane surface making it hydrophobic, but also the interior of the membrane pores; this was confirmed by the SEM–energy-dispersive X-ray spectroscopy (EDX) data of the cross section. Using commercially available flat polyethersulfone (PES) membrane (MicroPES 2F), Eykens et al. [32] explored a variety of coating techniques including vacuum plasma technology and atmospheric plasma technique to develop efficient hydrophobic–hydrophilic structures for successful DCMD. The CF_4 plasma treatment technique to develop a hydrophobic layer on existing asymmetric hydrophilic polyethersulfone membranes in both flat and hollow fiber forms was implemented by Wei et al. [33] for DCMD studies. Interestingly, this process ended up completely hydrophobizing the substrate including the back surface.

A major factor in such plasma treatment involves the extent of plasma exposure. Puranik et al. [34] studied the performance of flat hydrophilic PVDF membranes exposed to vacuum-based plasma polymerization depositing a polyfluorosiloxane coating as a function of treatment duration; by design, the plasma treatment duration was limited. The DCMD fluxes achieved were high. Here we report the DCMD performances of porous hydrophilic MicroPES hollow fibers of PES coated on the outside diameter (OD) with a plasma polymerized hydrophobic polyfluorosiloxane or polysiloxane coating. The performances of such coated hollow fiber membranes (HFMs) have been compared with hydrophobic PP based hollow fibers having a similar plasma polymerized coating. The gas gap in such PP-based hollow fibers through which water vapor diffuses is, however, very large compared to the very thin gas gap in the MicroPES hollow fibers with a plasma-polymerized coating, since the MicroPES hollow fiber is spontaneously wetted with water. Detailed performance results of a few flat hydrophilic porous PVDF membranes with a porous hydrophobic plasma polymerized coating of polyfluorosiloxane (not reported in [34]) are also provided here for comparison.

2. Experimental Materials and Methods

2.1. Materials and Chemicals

Hydrophilic PES hollow fibers of MicroPES type obtained from Membrana (Charlotte, NC, USA) were coated with an ultrathin plasma polymerized layer of hydrophobic polyfluorosiloxane or polysiloxane polymer on the outside surface. These fibers were asymmetric in structure with finer pores on the OD where plasma modification was done. Porous hydrophobic PP hollow fibers (PP 150/330) were also acquired from Membrana (Charlotte, NC, USA) and coated on the OD with an ultrathin plasma polymerized layer of hydrophobic polysiloxane or polyfluorosiloxane polymer using a continuous plasma process operated at 13.56 MHz. The monomer feed consisted of siloxane or blend of siloxane and perfluorohydrocarbon monomers selected from a group of 1,1,3,3 tetramethyldisiloxane (TMDSO), 1,3,5,7 tetramethylcyclotetrasiloxane (TMCTS), hexafluoroethane (HFE), perfluorohexane (PFHX), or perfluorooctane (PFOC). The monomer flow rate varied from 10–40 SCCM, the residence time in the reactor varied from 15 to 30 s and system pressure from 20–100 mTorr. The discharge power varied from 10–150 W. The basic properties of the uncoated porous hollow fiber membranes and their dimensions are provided in Table 1. Porous hydrophilic flat membranes of PVDF were obtained from MilliporeSigma (Bedford, MA, USA) and Pall Corp (Port Washington, NY, USA). The relevant properties of these two membranes are also listed in Table 1. These flat membranes were used as substrates for plasma coating. A highly porous ultrathin hydrophobic coating of polyfluorosiloxane polymer was deposited on these membranes by vacuum-based plasma polymerization [34].

Table 1. Properties of base hollow fiber membranes and flat membranes.

Base Membranes	d_m (μm) Mean Pore Size	ε_m (%) Porosity	δ_m (μm) Membrane Thickness	Internal Diameter (Hollow Fiber) (μm)
MicroPES hollow fiber (hydrophilic-polysulfone)	0.2	NA *	100 (wall thickness)	300
Polypropylene hollow fiber (hydrophobic)	>0.2	65	150 (wall thickness)	330
PVDF-MilliporeSigma (HVLP) (hydrophilic)	0.45	70	125	-
PVDF-Pall (hydrophilic)	0.1	82	72	-

* Not available.

Puranik et al. [34] have provided details of the plasma exposure variation between different samples of flat PVDF substrate membranes. The fluorosiloxane polymers on flat films were deposited in a tubular batch reactor under reaction conditions similar to those used for the HFMs, except that the reaction times were slightly longer. The samples AKS-6591-1M, AKS-6591-2M underwent plasma exposure for 1 min at two different positions in the reactor. The suffix -1 refers to a pre-electrode reactor position and the suffix -2 refers to the post-electrode reactor position. It is expected that position -2 would produce more cross-linked polymer than reactor position -1. Samples AKS-6592-1M, AKS-6593-1M, AKS-6594-1M were prepared by treating substrates for 2, 3, and 6 min respectively with plasma in the pre-electrode region.

2.2. Hollow Fiber Membrane Mini-Modules

A large number of mini-modules were built using both PP and PES hollow fiber membranes whose surfaces were modified by plasma polymerization. Each module contained 20 HFMs; each HFM was ~18 cm long. The total membrane surface area for each PP hollow fiber-based module was 7.12×10^{-3} m^2 whereas that for each PES hollow fiber-based module was 5.65×10^{-3} m^2 (area is based on the OD of the HFMs). The cylindrical shell of each module had an internal diameter (ID) of 0.95 cm and was made of polycarbonate. The HFM modules were regular parallel flow without rectangular crossflow [12–14] or radial crossflow arrangements [35] used in commercial scale modules to facilitate fouling-resistance and higher transfer coefficients/fluxes. The empty module shell volume was ~15.32 cm^3. The HFM packing density for PP HFs in the module was 8.78% and that for MicroPES HFs was 5.17%.

2.3. Methods and Procedures

The schematic for the experimental set up for studying the DCMD performances of plasma-coated hollow fiber membrane mini-modules is shown in Figure 1. It has two parallel flow loops: one for the brine to flow on the shell side of the membranes and the second loop for the distillate to flow through the lumen of the same membranes. In the membrane module, the brine stream and the distillate stream flow counter-currently. All flow loop components (e.g., instruments, pumps, sensors, tubing, valves, etc.) are made of materials that do not rust or contaminate the system.

Heated feed brine was stored in a polypropylene tank; its temperature was maintained by using a heat/temperature controller. It was pumped out into the shell side of the hollow fiber membrane module and then returned back to the tank. The hot brine inlet and outlet temperatures in the membrane module were measured and recorded using a temperature sensor. In the distillate loop, the distillate tank, filled with a certain minimum quantity of clean distilled water, was kept over an electronic weighing balance. Water from this tank was pumped through a heat exchanger into the module through the hollow fiber lumens counter-currently. The temperatures of the distilled water at the inlet and the outlet of the module were also measured using temperature sensors.

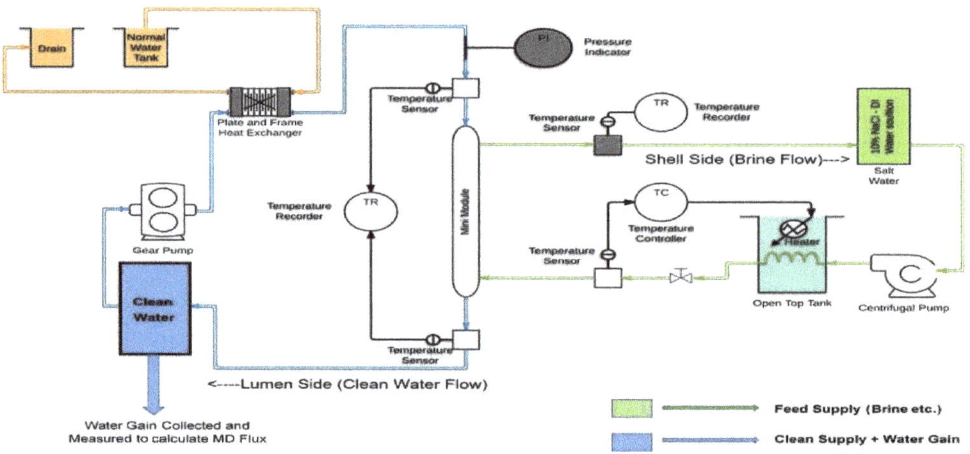

Figure 1. Experimental direct contact membrane distillation (DCMD) set up for studying the desalination performances of plasma coated hollow fiber membrane mini-modules.

The distillate stream exiting the module was returned to the clean water tank. Any weight change in the tank was continually monitored and recorded. Simultaneously, the salinity/conductivity of the distillate was recorded continually using a probe. The feed NaCl concentrations were: 1 wt%, 10 wt%, and 20 wt%; feed brine temperature was varied between 43.3 °C (110 °F) to 70 °C (158 °F). The incoming distillate side cold water temperature varied from 22.8 °C to 29.4 °C (73 °F to 85 °F) depending upon the heat exchanger, room temperature and, more importantly, the inlet brine temperature and the membrane distillation rate.

The DCMD experimental setup for flat membranes has been described in references [11,12,23,34]. A chlorinated polyvinyl chlroide (CPVC) based rectangular cell described in [23] with a membrane area of 0.0011 m^2 was used for the flat membranes. Studies with flat membranes used 1% NaCl solution. In most experiments, the incoming hot brine temperature varied between 60 °C and 85 °C. The distillate-in temperature was generally around 20 °C. The distillate production rate was obtained by the overflow rate from the distillate tank by a weighing machine and is obtained from Equation (1):

$$N_v\left(\frac{\text{kg}}{\text{m}^2.\text{h}}\right) = \frac{Increase\ in\ weight\ of\ water\ \left(=volume\ of\ water\ transferred(\text{L}) * water\ density\left(\frac{\text{kg}}{\text{L}}\right)\right)}{membrane\ area(\text{m}^2) * time(\text{h})} \tag{1}$$

To check for any salt leakage to the distillate side, the distillate side conductivity was measured using a conductivity meter (Orion 115A+, Thermo Fisher Scientific, Waltham, MA, USA). Any experiment under given conditions was run for around 3 h after steady state was reached.

Scanning electron microscopy was carried out using JSM 7900F Field Emission SEM (JEOL USA, Peabody, MA, USA) to develop micrographs of the surfaces as well as the cross section of the plasma polymerized coated hollow fiber membrane substrates of both PES and PP. The elements in the plasma-coated region determined via energy-dispersive X-ray spectroscopy (EDX) were reported in earlier studies [34]. Similar Si/F monomers and reaction conditions were used in this series of coatings on HFMs and therefore we expect the composition of these polymers to be similar. Note: due to the unique mechanism of the plasma polymerization process, plasma polymers can have significantly different elemental composition than the monomer used.

For liquid entry pressure (LEP) measurements, distilled water at room temperature was used as essentially stationary feed liquid on the shell side. A specially designed and fabricated LEP Test Apparatus (Applied Membrane Technology Inc., Minnetonka, MN, USA) was used to measure the "Leak Rate" of coated and uncoated membranes at different pressures. No water was present on the permeate side. The pressure settings used for coated PES membranes for leak measurement were 27.7, 48.3, and 68.9 kPag (4, 7 and 10 psig). The pressure of the feed distilled water was held constant for 30 min. If no permeate flow was observed, then the pressure was raised by 20.6 kPa (3 psi) and held constant until the final setting of 68.9 kPag (10 psig) was reached. The pressure at which we observed any permeation of water was considered to be the LEP. The permeation rate was measured at the LEP. For uncoated PES membranes, only 20.6 kPag (3 psig) pressure setting was used due to high leak rate: the hold time could not go beyond 10 min. For coated PP HFM modules, the same pressure settings were used but the hold time ranged from 30 to 60 min at each pressure setting.

3. Results

A total of 61 PP mini-modules and 24 PES mini-modules were tested for DCMD performance. We report here results of the 6900 series modules built with coated PES hollow fibers. The results from coated PP hollow fiber-based modules with the designations 6700 and 6800 are also being reported. Over 6000 h of DCMD data were collected for both HFM substrates having different porous hydrophobic polyfluorosiloxane/polysiloxane coatings at various feed brine concentrations and temperatures. Feed brine flow rate was ~2.66 L/min; the distillate flow rate was ~ 80 mL/min. Exiting hot brine temperature decreased from that at the inlet by ~0.5–3.3 °C (1–6 °F).

The average DCMD water vapor flux obtained from different PES substrate-based modules are illustrated in Table 2 for feed brine at 70 °C; two brine feed concentrations, 1% and 10%, were used. Of the different modules in this series, modules 6935, 6939 and 6940 yielded high water vapor fluxes between 26.4 and 27.6 kg/m^2-h with 1% feed brine. Since the feed temperature was 70 °C and these were parallel flow modules, we can reasonably expect significant flux enhancement if we had the hot brine in cross flow. For 10% brine feed at 70 °C, a lower but decent DCMD flux level of 17–19.5 kg/m^2-h was obtained for this series of coatings at 70 °C. These are high values. We deliberate on these flux values later.

Table 2. Average water vapor flux in DCMD for different polyethersulfone (PES)-based coated hollow fiber modules at two salt concentrations in feed at 70 °C.

Brine Concentration (%)	Module Series	Module Number	Type of Coating	Monomers Used	* W/FM	Flux (kg/m^2-h)
1	6900	6935	Polysiloxane	TMDSO	0.024	27.6
		6939	Polyfluorosiloxane	TMCTS/HFE	0.018	26.4
		6940	Polysiloxane	TMDSO	0.030	26.7
10	6900	6924	Polysiloxane	TMDSO	0.037	16.5
		6932	Polyfluorosiloxane	TMDSO/HFE	0.025	15.3
		6933	Polyfluorosiloxane	TMDSO/HFE	0.019	19.5
		6935	Polysiloxane	TMDSO	0.024	17.0
		6939	Polyfluorosiloxane	TMDSO/HFE	0.018	17.1

* W/FM—W is the power used in watts; F is the monomer feed rate in SCCM; M is the molecular weight of the monomer or blend.

The variable quantity, W/FM, is a composite plasma parameter which determines the average energy used in the plasma process per unit weight of monomer. An increase in the W/FM parameter led to a decrease in distillate flux at both salt concentrations (6935 vs. 6940 and 6933 vs. 6922). It has been demonstrated earlier by Sharma and Yasuda [36,37] that an increase in W/FM changed the kinetics of plasma polymerization and led to an increase in polymer deposition rate or an increase in crosslink density or both. Thus, an increase in W/FM can lead to an increase in the thickness or crosslink density of the coating resulting in a decrease in the distillate flux. Addition of fluoro monomer in the feed made the coating more hydrophobic and even less wettable. It also improved longevity of the membrane although the distillate flux marginally decreased. Halogen based compounds are known to act as catalyst and accelerate the rate of plasma polymer deposition [38]. This is perhaps why the polyfluorosiloxane coating in 6932 led to lower flux compared to the polysiloxane coating in 6935 in spite of similar plasma energy conditions. The fluoro monomer used in 6932 not just copolymerized but also resulted in thicker coating, which further reduced the pore size of the underlying substrate and increased the hydrophobic gap thickness, leading to reduction in the flux. Monomer TMCTS in 6939 also resulted in thicker coating and lower distillate flux due to higher O/Si ratio in the monomer [39].

Figure 2a–d, shown below, focus on the SEM pictures of the surfaces and cross sections of two candidate hollow fiber membranes in this series with a PES-based HFM substrate. The relevant module numbers from which the hollow fibers were taken out for SEM studies from this series are 6939 and 6940. Figure 2a,c describe respectively the surface and cross section of the 6939 series; the coating on the surface over the porous structure is clearly visible in both figures. Figure 2b,d provide the corresponding views of the 6940 series. From both Figure 2c,d, it is clear that the coating thicknesses are less than 0.4–0.5 µm. The coated pores appear to have a significant size distribution; further, the pores are somewhat elongated and elliptical in nature.

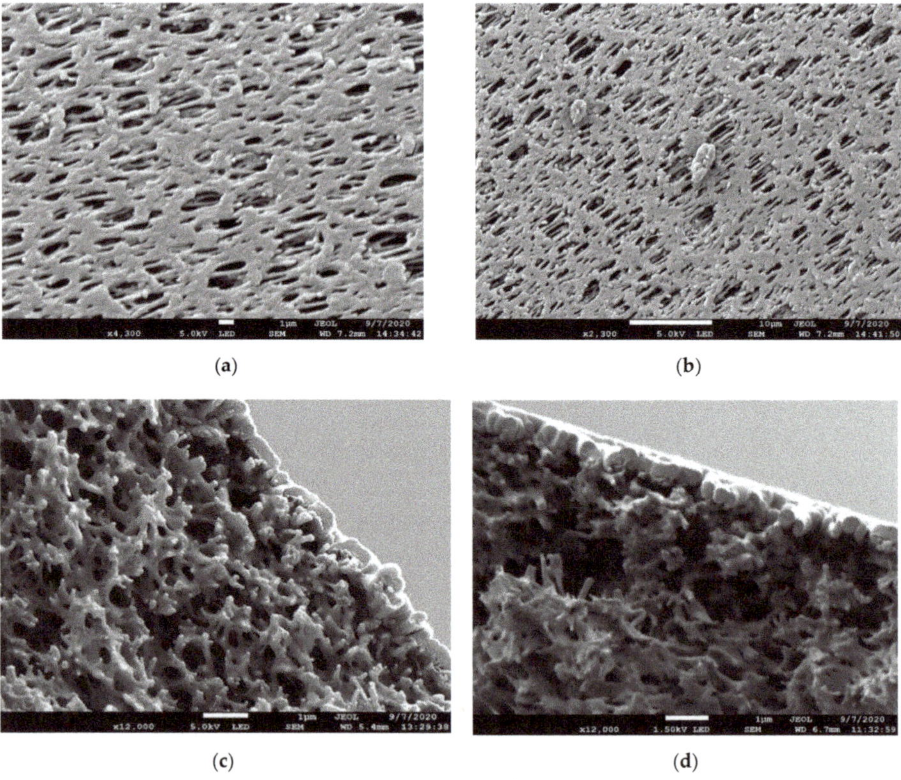

Figure 2. (**a**) SEM picture of the surface of 6939 coating on PES hollow fiber at 4300× magnification. (**b**) SEM picture of the surface of 6940 coating on PES hollow fiber at 2300× magnification. (**c**) SEM picture of the cross section of 6939 coating on PES hollow fiber at 12,000× magnification. (**d**) SEM picture of the cross section of 6940 coating on PES hollow fiber at 12,000× magnification.

Table 3 provides the average water vapor flux values obtained in DCMD with coated PP hollow fiber modules exposed to a 70 °C feed brine at two salt concentrations, 10% and 20%. Two coating series, designated 6700, 6800, were tested. The coating designated 6826 in the 6800 series has the highest flux of 15.5 kg/m^2-h for a 10 wt% brine feed. The values from other coatings are not too far off. Here again, higher brine concentration (lower vapor pressure) gave lower flux (coating in 6810 vs. 6713) and coating produced at lower W/FM or reduced reaction time offered higher flux in general (coating in 6713 vs. 6712 and coating in 6805 vs. 6804). It has been demonstrated by Sharma et al. [40] that a decrease in reaction time leads to a proportional decrease in polymer coating thickness. The reduced coating thickness is likely to have a lesser decrease in the pore size of the membrane, which would provide higher flux; the gas gap thickness is also reduced. Several different siloxane and fluoro monomers were used for these experiments. It was noted that the polyfluorosiloxane coatings based on PFHX and PFOC, especially the latter, had better film forming properties than HFE due to higher molecular weight and lower F/C ratio and produced thicker coatings and, hence, lower flux. Coating 6826, which was produced by reducing the amount of siloxane monomer in the mix by 50%, resulted in a significant increase in distillate flux. It is because the siloxane monomers, in general, are better film formers than the F-monomers and a reduction in silicone monomer, results in a thinner polymer coating and correspondingly higher flux. Plasma polymer coatings for this series were deposited at relatively lower W/FM because the polypropylene HFM (PP 150/330) could not tolerate high W/FM values especially in presence of F-monomers and became mechanically weak at high-energy conditions. The SEM-based micrographs of the surface and cross section of coating 6824 in the 6800 series are shown in Figure 3a,b, respectively.

Table 3. Average water vapor flux in DCMD for different polypropylene (PP)-based coated hollow fiber modules at two feed salt concentrations at 70 °C.

Brine Concentration (%)	Module Series	Module Number	Type of Coating	Monomers Used	W/FM	Flux (kg/m^2·h)
10	6700	6711	Polyfluorosiloxane	TMCTS/HFE	0.006	11.8
		6712	Polyfluorosiloxane	TMDSO/PFHX	0.008	10.1
		6713	Polyfluorosiloxane	TMDSO/PFHX	0.006	14.3
		6714	Polysiloxane	TMDSO	0.035	11.8
	6800	6823	Polyfluorosiloxane	TMDSO/PFHX	0.006	12.6
		6824	Polyfluorosiloxane	TMDSO/PFOC	0.006	11.4
		6825	Polyfluorosiloxane	TMDSO/PFOC	0.005	12.9
		6826	Polyfluorosiloxane	TMCTS */PFOC	0.006	15.5
20	6800	6803	Polyfluorosiloxane	TMDSO/PFHX	0.003	9.1
		6804	Polyfluorosiloxane	TMDSO/PFHX	0.007	7.6
		6805	Polyfluorosiloxane **	TMDSO/PFHX	0.007	10.1
		6806	Polyfluorosiloxane	TMCTS/PFHX	0.007	11.0
		6807	Polyfluorosiloxane **	TMCTS/PFHX	0.007	11.2
		6808	Polyfluorosiloxane	TMCTS/PFHX	0.007	10.1
		6810	Polyfluorosiloxane	TMDSO/PFHX	0.007	8.6

* Coating produced by reducing the amount of siloxane monomer by 50%; ** Reduced reaction time

The diameters of the coated surface pores in Figure 2a,b vary; a few are somewhat large. As a result, it is expected that the LEPs will be on the low side. Table 4 provides the LEP data and the corresponding leak rates using distilled water for a few types of coated PES hollow fibers. The leak rate for PP hollow fiber is provided for reference only. The LEP value of 48.3 kPag (7 psig) is on the low side. This LEP value is very close to that of flat hydrophilic PVDF substrate (MilliporeSigma; 0.4 μm pore size) having a very light coating, AKS-6591-1M [34]. However, the leak rates are extremely low. By manipulating plasma conditions, the LEP value can be substantially increased as was observed earlier with flat membranes [34]. In Table 5, we provide an idea about the rate of salt leakage as measured by change in conductivity of the distillate at the LEP for PP based hollow fiber membranes. Notice the lower salt leak rate in the polyfluorosiloxane coated PP hollow fiber coating.

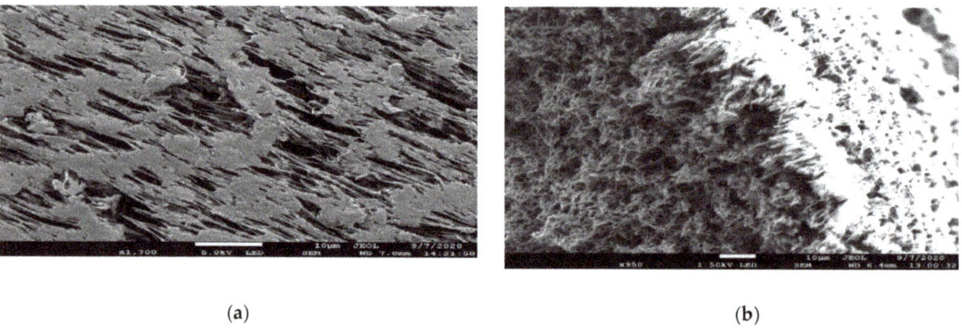

(a) (b)

Figure 3. (**a**) Surface of the 6824 coated PP hollow fiber (1700× magnification). (**b**) Cross section of the 6824 coated PP hollow fiber (950× magnification).

Table 4. Data on liquid entry pressure (LEP) and leak rate in hollow fiber mini-modules * having selected coatings.

Hollow Fiber Membrane (HFM) Type	LEP Value in kPag (psig) for the Leak Rate Measured	Leak Rate, mL/h
Coated PES HFM, 6935	48.3 (7)	0.2
Coated PES HFM, 6939	48.3 (7)	1.1
Coated PES HFM, 6940	48.3 (7)	0.4
Uncoated PES HFM	48.3 (7)	4211
Coated PP HFM	68.9 (10)	0.2

* Each mini-module has 20 HFMs, 18 cm in length.

Table 5. Salt leak rate of polysiloxane and polyfluorosiloxane coating for PP based hollow fiber membrane at 70 °C.

Brine Concentration	Module/Coating No.	Coating Type	Average Water Flux ($kg/m^2 \cdot h$)	Conductivity Change (4 h)
20%	6714	Polysiloxane	11.4	0.1–5.8 µs/cm
20%	6806	Polyfluorosiloxane	10.0	0.2–1.6 µs/cm

In the earlier study using flat hydrophilic PVDF membranes [34], we had observed a very limited flux reduction accompanied by a considerable increase in LEP when the plasma polymerization was just enhanced a bit. For example, flat membrane samples AKS-6591-1M, AKS-6591-2M underwent plasma exposure for 1 min at two different positions in the reactor with more crosslinking at -2 position due to a more intensive plasma exposure; the corresponding LEP values for the two coatings were 62 and 110 kPag (9 and 16 psig) respectively. However, the flux for the membrane modified at -2 position was reduced by only 2 kg/m^2-h. We expect an almost similar behavior for hollow fiber membranes since the plasma polymerization conditions were very similar. This happens during earlier stages of coating. Figure 6 in reference [34] describes the experimental data for DCMD flux vs. plasma polymerization time.

A comparison of the flux performances of the two types of hollow fiber substrate is useful since in the hydrophilic PES hollow fibers, the length of the water vapor diffusion path is very small. It essentially spans the length of the porous plasma polymerized hydrophobic coating and tends toward creating conditions for orifice flow. Therefore, the flux level is likely to be higher; however, conductive heat loss is increased, which tends to reduce the flux [23]. In the PP substrate-based coated hollow fibers, the vapor diffusion path is much longer. The fluxes are going to be lower but conductive heat loss will also be significantly lower [23]. The overall performances of the two types of coated HFMs are shown in Figures 4 and 5 for two different salt concentrations, 1 wt% and 10 wt%, respectively.

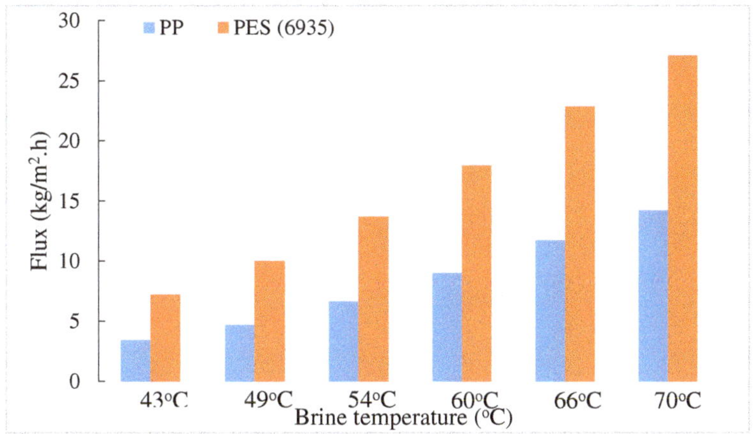

Figure 4. Comparison of average water vapor flux between PP (6713) and PES (6935) based coated hollow fiber modules for 1% brine feed at different temperatures.

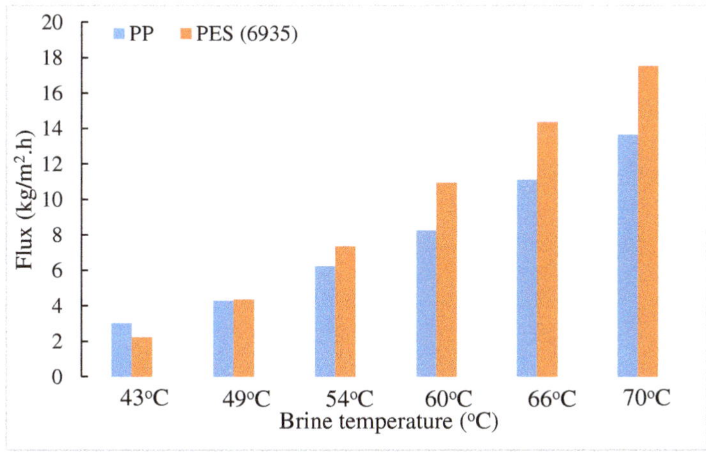

Figure 5. Comparison of average water vapor flux between PP (6713) and PES (6935) based coated hollow fiber modules for 10% brine feed at different temperatures.

It appears that the flux levels in coated PES hollow fibers are always higher than those in coated PP hollow fibers for 1 wt% brine feed by a substantial amount. On the other hand, that difference is significantly smaller in the case of a 10% brine feed along with a generally lower level of flux.

Since the HFM packing density for PP HFs in the module was 8.78% and that for MicroPES HFs was 5.17% for a constant module volume, the open volume was significantly larger in the PES HFM modules. The extent of bypassing/channeling of hot brine would be larger in the PES HFM modules. With 1% hot feed brine, the extent of temperature and concentration polarization influenced by this bypassing and therefore water vapor pressure reduction would be much less. On the other hand, the same level of bypassing/channeling would have much larger effect on concentration/temperature polarization with 10% hot feed brine in PES modules and correspondingly larger vapor pressure reduction. Hence, we observe a significantly larger flux reduction with PES modules.

Further, this difference between the two sets of HFs keeps on decreasing as the brine feed temperature decreases with 10% brine feed. There is almost no difference at a feed temperature of 49 °C. The relative flux levels are reversed at a low feed temperature of 43 °C. This is likely due to an enhanced contribution of conductive flux loss vis-à-vis the total heat flux across the membrane (which is the sum of convective heat flux due to vapor transport and the conductive heat flux) for the coated PES hollow fiber membranes. This is aided by more bypassing in the PES modules.

The modules with coated PP hollow fibers, when tested on 1% brine feed, showed very little salt leakage, even after running for more than 1000 h. The distillate conductivity increased only by a few μS/cm. When testing PP hollow fibers on 10% brine feed, the salt leakage increased just a little bit more. Coated HFMs based on MicroPES substrate, particularly 6935, performed well with 1% brine; the very low leakage rate was comparable to that of PP HFMs. For a feed of 10% salt, the leakage was slightly higher, suggesting the need for a longer plasma polymerization time or a higher W/FM to prevent any salt leakage at all.

To provide a comparative perspective on the water vapor fluxes from these coated HFMs vis-à-vis those obtained with flat membranes, we now briefly describe the observed flux behaviors of a few porous flat hydrophilic PVDF membranes having a thin plasma polymerized polyfluorosiloxane coating for 1 wt% salt containing feed. The general characteristics of these membranes in terms of LEP values and overall relative flux behavior vis-à-vis the plasma polymerization time are available in the earlier study [34]. However, detailed flux behavior vis-à-vis feed temperature and flow rates were not available for the specific membranes considered here.

We show how the water vapor flux of the flat coated membranes varies with feed brine temperature and brine flow rate respectively in Figure 6a,b. The plasma polymerization levels for these membranes and therefore the coating thickness were just a bit higher than that of AKS-6591-1M and AKS-6591-2M membranes, which had the lowest plasma polymerization time, the lowest LEP value and the highest water vapor fluxes. We have already mentioned that the LEP values of these membranes (AKS-6591-1M and AKS-6591-2M) were comparable to those of the coated MicroPES HFMs of this study. Figure 6b shows how strongly the water vapor flux decreases at 70 °C with a decrease in feed brine flow rate and increase in temperature polarization. The cell design for flat membranes creates a significant amount of flow mixing to reduce temperature polarization. That is one of the reasons why the water vapor flux at 70 °C with the PES based hollow fiber modules in Figure 4 are smaller than those of flat membranes (Figure 6b).

Figure 7a,b illustrate similar data for coated flat hydrophilic PVDF membranes designated AKS-6594-1 and AKS-6594-2 with Millipore substrate. The coatings were generated with a longer treatment time and hence led to lower fluxes. These coatings also show significantly higher LEP values [34]. In fact, one can now produce with reasonable certainty a certain type of plasma polymerized coating on a given substrate that will yield a certain small range of LEP and water vapor flux level, with due considerations for feed and distillate temperatures, and feed and distillate side fluid mechanical conditions.

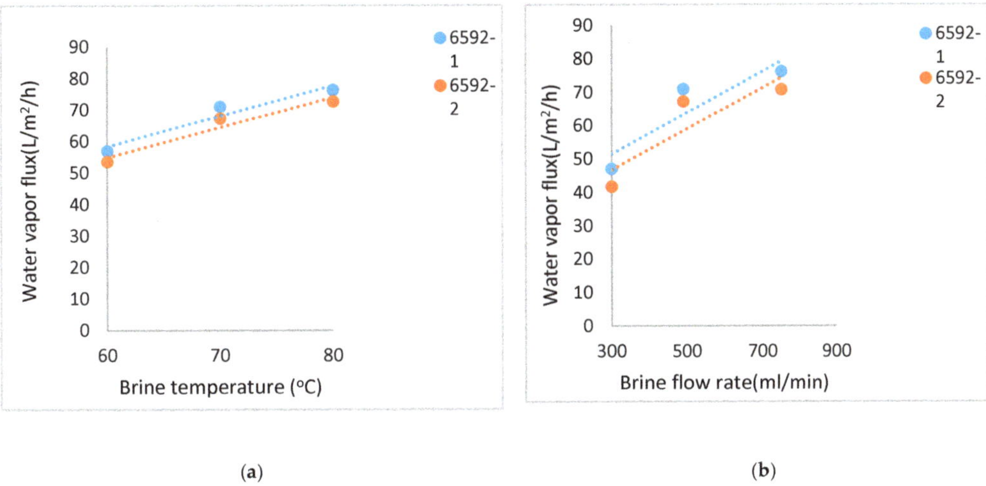

(**a**) (**b**)

Figure 6. (**a**) Water vapor flux values for different brine temperatures and a constant brine flow rate of 490 mL/min for AKS-6592-1 and AKS- 6592-2 Millipore. (**b**) Water vapor flux at various brine flow rates and a constant brine inlet temperature of 70 °C for AKS- 6592-1 and AKS- 6592-2 Millipore.

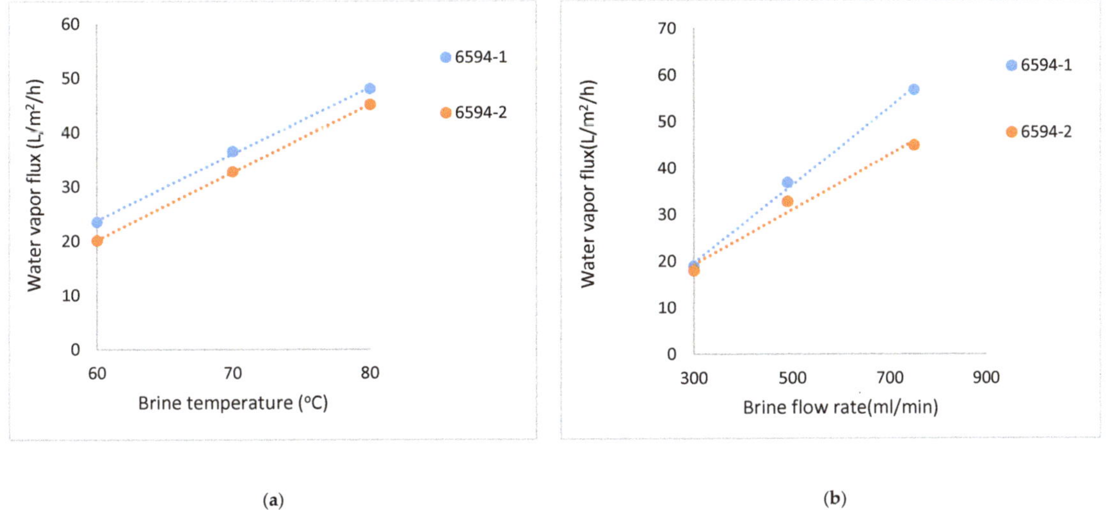

(**a**) (**b**)

Figure 7. (**a**) Water vapor flux values for different brine temperatures and a constant brine flow rate of 490 mL/min for AKS-6594-1 and AKS- 6594-2 Millipore. (**b**) Water vapor flux at various brine flow rates and a constant brine inlet temperature of 70 °C for AKS- 6594-1 and AKS- 6594-2 Millipore.

4. Discussion

It is useful to deliberate on the volumetric distilled water production rate from the types of HFM-based modules used. Consider the distilled water flux from PES substrate-based modules in the 6900 series shown in Table 2. An average value of 27 kg/m^2-h may be assumed from a module having a membrane surface area of 5.6 × 10^{-3} m^2 and an internal volume of 15.32 cm^3. The corresponding volumetric productivity is 9869 kg/m^3-h. Considering that the volume fraction occupied by HFMs in these modules was only 5.17%, we can assume that this value can be easily tripled (or quadrupled) leading to a potential volumetric distilled water production rate of ~30,000 kg/m^3-h. Such an estimate is useful for estimating the expected performance level from hollow fiber modules. It is considerably larger than modules based on flat membranes, spiral-wound or otherwise. Crossflow hollow fiber membrane modules designed and used earlier in DCMD can have a HFM packing fraction of $\geq$0.2 [35] in the cylindrical module. We recommend keeping the HFM packing volume fraction to $\leq$0.2 in view of considerations discussed in the next paragraph.

A related issue involves the pressure drop experienced by the hot brine flowing on the shell side of a HFM-based module. Since the hollow fiber bed has a relatively open cross-section, the pressure drop experienced by hot brine during passage through one module with a relatively deep fiber bed [13] is around ~13.6 kPa (2 psi). However, since the fractional water recovery in DCMD is not very high, a cascade of stages may be needed for high water recovery [19]. If at each stage, a pump is used to drive the brine, the pressure of the brine in any module will rarely exceed 10–15 kPag. Therefore, the LEP value of the current HFMs will not be exceeded. However, if reduced numbers of pumps are used, the hot brine pressure will be higher at the pump exit; that would require a higher value of LEP for coated MicroPES HFMs and a small reduction in achievable water vapor flux. These and other considerations on module design and system improvement of mem-brane distillation identified in [41] are of importance in further developments being undertaken.

The internal mechanism of water vapor flux in hydrophilic HFMs having a hydrophobic plasma polymerized coating in DCMD depends on water vapor diffusion through a thin gas-filled gap with colder water at the end of the thin gap. A transport model for DCMD in such a configuration has to take into account a number of items, which are unknown at this time: the thickness of the gas gap and the shape of its boundaries influenced by plasma coating; the thermal conductive resistance in parallel through the hydrophilic substrate membrane, as well as that through the water-filled pores of the substrate at the end of which distilled water flows. The temperature gradient in the water-filled pores of the substrate membrane is important as well since hot water vapor is condensing at the gas-liquid boundary inside the hydrophilic water-filled membrane pores while cold distilled water flows at the other end. Further, one has to be able to account for temperature polarization in the hot feed brine. This configuration thus requires an extensive investigation.

An aspect of plasma polymerization-based membrane surface modification is the general absence of the use of nanoparticles and organic solvents resulting in generation of toxic waste. Many advanced membranes being developed in MD involve production of such wastes. A recent example of MD membrane development that avoids/reduces such waste production is provided in [42].

5. Concluding Remarks

A very thin porous hydrophobic coating on a porous hydrophilic membrane can provide a membrane configuration for enhanced water vapor flux in DCMD with the hydrophilic substrate wetted by the cold distilled water stream. We demonstrated here that such a configuration can be achieved successfully with hydrophilic porous MicroPES hollow fiber membranes of polyethersulfone coated on the outside surface with a 0.4–0.5 μm thick porous hydrophobic polyfluorosiloxane coating using a vacuum-based plasma polymerization process. The HFM surface area in the module based on 20 HFMs was 5.6×10^{-3} m^2. The water vapor flux level achieved with such hollow fiber membrane-based modules from a 1 wt% NaCl-containing brine feed at 70 °C was in the range of 26.4 to 27.6 kg/m^2-h. These flux values were found to be significantly higher than those from hollow fiber membrane modules built using porous hydrophobic hollow fiber membranes of PP having a similar plasma polymerized coating; such PP-based hollow fibers require water vapor to traverse a long gas gap in the porous hydrophobic substrate. The water vapor fluxes from MicroPES hollow fibers are reduced to ~17–19 kg/m^2-h when exposed to a 70 °C feed of 10 wt% of salt solution. The salt leakage was extremely low in both cases.

The calculated values of the volumetric productivity of distilled water from such modules are substantial.

Author Contributions: Conceptualization, S.P.C., A.K.S. and K.K.S.; funding acquisition, S.P.C. and K.K.S.; supervision, S.P.C., K.K.S.; methodology, S.P.C. and K.K.S.; plasma polymerization-based coating, A.K.S.; hollow fiber module making, S.S.; hollow fiber data gathering, A.J., C.C., S.S.; flat membrane characterization and data gathering, A.A.P., L.R.; SEM characterization, J.C.; hollow fiber data analysis, S.P.C., A.K.S.; K.K.S., J.C.; writing, review and editing—K.K.S., S.P.C., A.K.S. All authors have read and agreed to the published version of the manuscript.

Funding: This research received no external funding.

Acknowledgments: Aishwarya Puranik carried out desalination experiments for her MS Thesis at the Otto York Department of Chemical, Biological, and Pharmaceutical Engineering, New Jersey Institute of Technology. Lydia Rodrigues and John Chau gratefully acknowledge support for this research from the NSF Industry/University Cooperative Research Center for Membrane Science, Engineering, and Technology that has been supported via NSF Award IIP 1034710. We acknowledge both MilliporeSigma and Pall Corporation for providing the PVDF membranes.

Conflicts of Interest: The authors describe no conflict of interest.

References

1. Yang, X.; Fane, A.G.; Wang, R. Membrane Distillation: Now and Future. In *Desalination*, 2nd ed.; Kucera, J., Ed.; Scrivener Publishing LLC: Beverly, MA, USA, 2018; pp. 329–386.
2. Drioli, E.; Ali, A.A.A.; Macedonio, F. Membrane distillation: Recent developments and perspectives. *Desalination* **2015**, *356*, 56–84. [CrossRef]
3. Deshmukh, A.; Boo, C.; Karanikola, V.; Lin, S.; Straub, A.P.; Tong, T.; Warsinger, D.M.; Elimelech, M. Membrane distillation at the water-energy nexus: Limits, opportunities, and challenges. *Energy Environ. Sci.* **2018**, *11*, 1177–1196. [CrossRef]
4. Camacho, L.M.; Dumée, L.F.; Zhang, J.; Li, J.-D.; Duke, M.; Gomez, J.D.; Gray, S. Advances in Membrane Distillation for Water Desalination and Purification Applications. *Water* **2013**, *5*, 94–196. [CrossRef]
5. Alkhudhiri, A.; Darwish, N.A.; Hilal, N. Membrane distillation: A comprehensive review. *Desalination* **2012**, *287*, 2–18. [CrossRef]
6. Lawson, K.W.; Lloyd, D.R. Membrane distillation. *J. Membr. Sci.* **1997**, *124*, 1–25. [CrossRef]
7. Sirkar, K.K. Other New Membrane Processes. In *Membrane Handbook*; Ho, W.S.W., Sirkar, K.K., Eds.; Van Nostrand Reinhold, 1992; Springer Nature: Berlin/Heidelberg, Germany, 2012; pp. 885–912.
8. Chen, H.-C.; Chen, Y.-R.; Yang, K.-H.; Yang, C.-P.; Tung, K.-L.; Lee, M.-J.; Shih, J.-H.; Liu, Y.-C. Effective reduction of water molecules' interaction for efficient water evaporation in desalination. *Desalination* **2018**, *436*, 91–97. [CrossRef]
9. Anvari, A.; Kekre, K.M.; Yancheshme, A.A.; Yao, Y.; Ronen, A. Membrane distillation of high salinity water by induction heated thermally conducting membranes. *J. Membr. Sci.* **2019**, *589*, 117253. [CrossRef]
10. Dudchenko, A.V.; Chen, C.; Cardenas, A.; Rolf, J.; Jassby, D. Frequency-dependent stability of CNT Joule heaters in ionizable media and desalination processes. *Nat. Nanotechnol.* **2017**, *12*, 557–563. [CrossRef]
11. Singh, D.; Prakash, P.; Sirkar, K.K. De-oiled Produced Water Treatment using Direct Contact Membrane Distillation. *I&EC Res.* **2013**, *52*, 13439–13448.
12. He, F.; Gilron, J.; Lee, H.; Song, L.; Sirkar, K. Potential for scaling by sparingly soluble salts in crossflow DCMD. *J. Membr. Sci.* **2008**, *311*, 68–80. [CrossRef]
13. Song, L.; Ma, Z.; Liao, X.; Kosaraju, P.B.; Irish, J.R.; Sirkar, K.K. Pilot plant studies of novel membranes and devices for direct contact membrane distillation-based desalination. *J. Membr. Sci.* **2008**, *323*, 257–270. [CrossRef]
14. He, F.; Sirkar, K.; Gilron, J. Effects of antiscalants to mitigate membrane scaling by direct contact membrane distillation. *J. Membr. Sci.* **2009**, *345*, 53–58. [CrossRef]
15. Wang, Z.; Chen, Y.; Sun, X.; Duddu, R.; Lin, S. Mechanism of pore wetting in membrane distillation with alcohol vs. surfactant. *J. Membr. Sci.* **2018**, *559*, 183–195. [CrossRef]
16. Chen, Y.; Tian, M.; Li, X.; Wang, Y.; An, A.K.; Fang, J.; He, T. Anti-wetting behavior of negatively charged superhydrophobic PVDF membranes in direct contact membrane distillation of emulsified wastewaters. *J. Membr. Sci.* **2017**, *535*, 230–238. [CrossRef]
17. Mostafa, M.G.; Zhu, B.; Cran, M.J.; Dow, N.; Milne, N.A.; Desai, D.; Duke, M. Membrane Distillation of Meat Industry Effluent with Hydrophilic Polyurethane Coated Polytetrafluoroethylene Membranes. *Membranes* **2017**, *7*, 55. [CrossRef] [PubMed]
18. Tang, L.; Iddya, A.; Zhu, X.; Dudchenko, A.V.; Duan, W.; Turchi, C.; Vanneste, J.; Cath, T.Y.; Jassby, D. Enhanced Flux and Electrochemical Cleaning of Silicate Scaling on Carbon Nanotube-Coated Membrane Distillation Membranes Treating Geothermal Brines. *ACS Appl. Mater. Interfaces* **2017**, *9*, 38594–38605. [CrossRef]
19. He, F.; Gilron, J.; Sirkar, K.K. High water recovery in direct contact membrane distillation using a series of cascades. *Desalination* **2013**, *323*, 48–54. [CrossRef]

20. Winter, D.; Koschikowski, J.; Gross, F.; Maucher, D.; Düver, D.; Jositz, M.; Mann, T.; Hagedorn, A. Comparative analysis of full-scale membrane distillation contactors—methods and modules. *J. Membr. Sci.* **2017**, *524*, 758–771. [CrossRef]
21. Singh, D.; Sirkar, K.K. Desalination by Airgap Membrane Distillation Using a Two Hollow-fiber-set Membrane Module. *J. Membr. Sci.* **2012**, *421–422*, 172–179. [CrossRef]
22. Schwantes, R.; Seger, J.; Bauer, L.; Winter, D.; Hogen, T.; Koschikowski, J.; Geißen, S.-U. Characterization and Assessment of a Novel Plate and Frame MD Module for Single Pass Wastewater Concentration–FEED Gap Air Gap Membrane Distillation. *Membranes* **2019**, *9*, 118. [CrossRef]
23. Li, L.; Sirkar, K.K. Influence of microporous membrane properties on the desalination performance in direct contact membrane distillation. *J. Membr. Sci.* **2016**, *513*, 280–293. [CrossRef]
24. Lee, H.; He, F.; Song, L.; Gilron, J.; Sirkar, K. Desalination with a cascade of cross-flow hollow fiber membrane distillation devices integrated with a heat exchanger. *AIChE J.* **2010**, *57*, 1780–1795. [CrossRef]
25. Khayet, M.; Mengual, J.; Matsuura, T. Porous Hydrophobic/hydrophilic Composite Membranes- Application in Desalination using Direct Contact Membrane Distillation. *J. Membr. Sci.* **2005**, *252*, 101–113. [CrossRef]
26. Khayet, M.; Matsuura, T.; Mengual, J.; Qtaishat, M. Design of novel direct contact membrane distillation membranes. *Desalination* **2006**, *192*, 105–111. [CrossRef]
27. Khayet, M.; Matsuura, T.; Qtaishat, M.R.; Mengual, J.I. Porous hydrophobic/hydrophilic composite membranes preparation and application in DCMD desalination at higher temperatures. *Desalination* **2006**, *199*, 180–181. [CrossRef]
28. Qtaishat, M.; Khayet, M.; Matsuura, T. Guidelines for preparation of higher flux hydrophobic/hydrophilic composite membranes for membrane distillation. *J. Membr. Sci.* **2009**, *329*, 193–200. [CrossRef]
29. Floros, I.N.; Kouvelos, E.P.; Pilatos, G.I.; Hadjigeorgiou, E.P.; Gotzias, A.D.; Favvas, E.P.; Sapalidis, A.A. Enhancement of Flux Performance in PTFE Membranes for Direct Contact Membrane Distillation. *Polymers* **2020**, *12*, 345. [CrossRef]
30. Liu, Y.; Xiao, T.; Bao, C.; Fu, Y.; Yang, X. Fabrication of novel Janus membrane by nonsolvent thermally induced phase separation (NTIPS) for enhanced performance in membrane distillation. *J. Membr. Sci.* **2018**, *563*, 298–308. [CrossRef]
31. Tian, M.; Yin, Y.; Yang, C.; Zhao, B.; Song, J.; Liu, J.; Li, X.-M.; He, T. CF4 plasma modified highly interconnective porous polysulfone membranes for direct contact membrane distillation (DCMD). *Desalination* **2015**, *369*, 105–114. [CrossRef]
32. Eykens, L.; De Sitter, K.; Dotremont, C.; Pinoy, L.; Van Der Bruggen, B. Coating techniques for membrane distillation: An experimental assessment. *Sep. Purif. Technol.* **2018**, *193*, 38–48. [CrossRef]
33. Wei, X.; Zhao, B.; Li, X.-M.; Wang, Z.; He, B.-Q.; He, T.; Jiang, B. CF4 plasma surface modification of asymmetric hydrophilic polyethersulfone membranes for direct contact membrane distillation. *J. Membr. Sci.* **2012**, 164–175. [CrossRef]
34. Puranik, A.A.; Rodrigues, L.N.; Chau, J.; Li, L.; Sirkar, K.K. Porous hydrophobic-hydrophilic composite membranes for direct contact membrane distillation. *J. Membr. Sci.* **2019**, *591*, 117225. [CrossRef]
35. Singh, D.; Li, L.; Obusckovic, G.; Chau, J.; Sirkar, K.K. Novel cylindrical cross-flow hollow fiber membrane module for direct contact membrane distillation-based desalination. *J. Membr. Sci.* **2018**, *545*, 312–322. [CrossRef]
36. Sharma, A.K.; Yasuda, H. Polymerization of methane. *J. Appl. Polym. Sci.* **1989**, *38*, 741–754. [CrossRef]
37. Sharma, A.K.; Yasuda, H. Plasma polymerization of tetramethyldisiloxane by a magnetron glow discharge. *Thin Solid Films* **1983**, *110*, 171–184. [CrossRef]
38. Sharma, A.K.; Millich, F.; Hellmuth, E.W. Propylene Glow Discharge Polymerization in the Presence of Bromotrichloromethane. *ACS Symp. Ser.* **1979**, 53–64. [CrossRef]
39. Sharma, A.K. Organosiloxane Films for Gas Separations. U.S. Patent 9,339,770 B2, 17 May 2016.
40. Sharma, A.K.; Millich, F.; Hellmuth, E.W. Adhesion and Hydrophobicity of Glow Discharge Polymerized Propylene Coating. *J. Appl. Phys.* **1978**, *49*, 5055–5059. [CrossRef]
41. Lu, K.J.; Chung, T.S. *Membrane Distillation: Membranes, Hybrid Systems and Pilot Studies*, 1st ed.; CRC Press: Boca Raton, FL, USA, 2020.
42. Lu, K.; Zhao, D.; Chen, Y.; Chang, J.; Chung, T.-S. Rheologically controlled design of nature-inspired superhydrophobic and self-cleaning membranes for clean water production. *npj Clean Water* **2020**, *3*, 1–10. [CrossRef]

Article

CFD Investigation of Spacer-Filled Channels for Membrane Distillation

Mariagiorgia La Cerva [1], Andrea Cipollina [1], Michele Ciofalo [1,*], Mohammed Albeirutty [2,3], Nedim Turkmen [3], Salah Bouguecha [3] and Giorgio Micale [1]

1 Dipartimento di Ingegneria, Università degli Studi di Palermo, viale delle Scienze ed.6, 90128 Palermo, Italy
2 Center of Excellence in Desalination Technology, King Abdulaziz University, Jeddah 21589, Saudi Arabia
3 Mechanical Engineering Department, King Abdulaziz University, Jeddah 21589, Saudi Arabia
* Correspondence: michele.ciofalo@unipa.it

Received: 11 June 2019; Accepted: 15 July 2019; Published: 25 July 2019

Abstract: The membrane distillation (MD) process for water desalination is affected by temperature polarization, which reduces the driving force and the efficiency of the process. To counteract this phenomenon, spacer-filled channels are used, which enhance mixing and heat transfer but also cause higher pressure drops. Therefore, in the design of MD modules, the choice of the spacer is crucial for process efficiency. In the present work, different overlapped spacers are investigated by computational fluid dynamics (CFD) and results are compared with experiments carried out with thermochromic liquid crystals (TLC). Results are reported for different flow attack angles and for Reynolds numbers (Re) ranging from ~200 to ~800. A good qualitative agreement between simulations and experiments can be observed for the areal distribution of the normalized heat transfer coefficient. Trends of the average heat transfer coefficient are reported as functions of Re for the geometries investigated, thus providing the basis for CFD-based correlations to be used in higher-scale process models.

Keywords: desalination; membrane distillation; spacer-filled channel; temperature polarization; computational fluid dynamics; thermochromic liquid crystals

1. Introduction

In regions where the existing freshwater sources do not meet the water demand, production of potable water from saline and brackish waters is needed [1]. Among the desalination processes, membrane distillation (MD) has recently attracted significant interest due to its features, which make MD a process easily scalable to small–medium size [2–4]. MD is a thermally driven membrane-based process where the transport of water vapor is realized through a micro-porous hydrophobic membrane, which leads to a theoretical 100% rejection of salts. The separation process is driven by the vapor pressure difference at the membrane/liquid interfaces due to the temperature gradient existing between the two sides of the membrane. MD is characterized by its compactness and low operational pressure requirement as well as low working temperature. These features make it possible to power MD by low grade heat (waste heat) or renewable energy and boost its applicability in remote areas and small scale production [5–7]. Another important distinct feature of MD is its ability to desalinate highly saline brines and wastewater. MD is the most appropriate option to increase the recovery ratio of multi-stage flash (MSF) and reverse osmosis (RO) plants due to it its ability to treat feeds with high salinity and impurities. One main advantage of MD technology is that evaporation and condensation surfaces are tightly packed, thus leading to a compact system with low capital cost per unit product. The classification of MD systems is related to the adopted condensation methods. In general, MD systems are classified into four basic simple configurations designed relative to the permeate vapor condensation method used. These include direct contact membrane distillation (DCMD), air gap membrane distillation (AGMD), sweeping gas membrane distillation (SGMD), and vacuum membrane distillation (VMD) [5]. The succession of these different configurations was dictated by a need to improve the performance of the process. DCMD is the simplest configuration that has liquid phases (hot feed and cold permeate) in direct contact with both sides of the porous hydrophobic membrane. The vapor diffusion path is limited to the thickness of the membrane, therefore reducing the mass and heat transfer resistances and thus enhancing the process [8,9].

Despite these advantages and the many experimental and theoretical studies conducted in recent years, MD is not yet fully developed from the commercial point of view because it is still highly energy demanding (typical thermal energy consumption ~100 kWh/m^3) compared, for example, to multiple effect distillation, or MED (thermal energy consumption ~40 kWh/m^3 for large plants) and reverse osmosis, or RO (mechanical energy consumption 2–4 kWh/m^3) [10–12]. Other reasons why MD is still on a pilot scale are related to membrane, module design, and low water flux. One of the main causes responsible for low water flux is the temperature polarization phenomenon, while the concentration polarization has a negligible effect on the permeate flow rate reduction [13]. Temperature polarization can lead to a dramatic reduction of the actual transmembrane temperature driving force [10,13]. A reduction of these effects and a performance improvement could be achieved by the optimization of the geometry of the channels and of the adopted spacers acting as mixing promoters.

Computational fluid dynamics (CFD) is suitable to investigate flow field and transport phenomena in spacer-filled channels. Since the 1990s, CFD has been used to study channels with spacers, by 2D [14–16] or 3D simulations [17–20], in the steady regime [16,21–24] or considering unsteady flows [25–27]. In the unsteady regime, very accurate direct numerical simulations (DNS) become computationally expensive, and turbulence models may be more suitable [28,29]. In regard to CFD results validation, in most of the literature only global quantities from CFD and experiments are compared, such as the friction coefficient and the average Nusselt or Sherwood numbers [25,30,31]. However, local data (i.e., flow fields and temperature distributions) are more suitable than global quantities because they better allow one to check if the assumptions made in CFD simulations (e.g., assuming simple boundary conditions or neglecting geometrical irregularities) reproduce sufficiently well the actual physics of the system. Experiments based on thermochromic liquid crystals (TLC) and digital image processing (DIP) can provide local results, such as temperature and heat transfer coefficient distributions, which can also be processed to obtain global quantities [27,32–35].

The aim of the present work is to characterize and design novel prototypes of spacers for membrane separation modules. We performed 3D simulations from the laminar to the unsteady regime by using, in this latter case, the shear stress transport (SST) turbulence model. For selected test cases, CFD results were compared with experimental data obtained in a purpose built test facility where the TLC technique was used. The validated CFD model was then used to investigate several spacers in different conditions.

2. Materials and Methods

2.1. Geometry, Operating Conditions, and Performance Parameters

An overlapped spacer of the kind examined in the present study is characterized by a few geometric parameters (Figure 1), namely: the filament diameter (d); the orthogonal distance between filaments (pitch, P); the intrinsic angle between filaments (θ); and the flow attack angle (α) between the directions of the main flow and of the filament layer adjacent to a specified (e.g., top) wall. Note that d and P are assumed to be the same for the two overlapped layers of filaments. In the present study, for the above parameters the values reported in Table 1 were investigated by both experiments and CFD simulations. In order to exactly reproduce the geometry of the experimental module, the geometrical model for CFD was created from the same CAD files used for the 3D printing of the spacer. The channel height H, chosen to be within the range of typical spacer thickness values in MD [18,36,37], was not exactly equal to 2d (twice the filament diameter), i.e., 4 mm, but slightly less (3.8 mm) because a certain amount of interpenetration existed. Thus, the pitch to channel height ratio was $P/H = 2.63$.

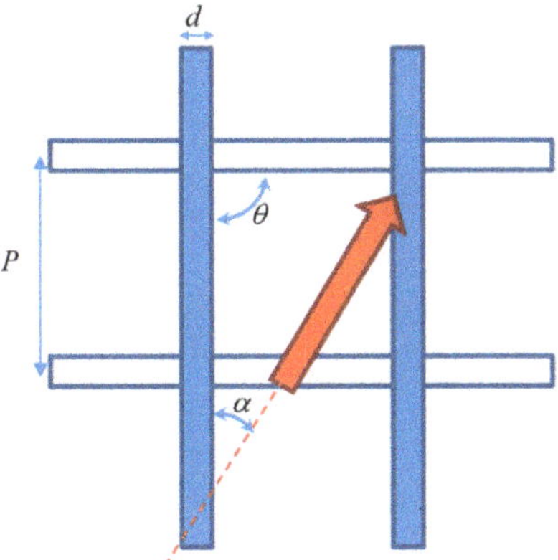

Figure 1. Sketch of a generic overlapped spacer with the main geometric parameters. The red arrow indicates the flow direction.

Table 1. Investigated values of geometric parameters of spacers.

d (mm)	P (mm)	θ (°)	α (°)
		30	15
		60	30
			0
			15
2	10		30
		90	45
			60
			75
			90

Both in the experiments and in the computations, the bottom wall was assumed adiabatic and heat transfer occurring only through the top one (active wall), in which distributions of temperature, wall heat flux, and heat transfer coefficient were assessed. This last quantity was locally defined as:

$$h = \frac{q''}{T_{bulk} - T_{wall}} \tag{1}$$

in which T_{bulk} is the local fluid bulk temperature, while q'' and T_{wall} are the local wall heat flux and wall temperature, respectively. For the reasons discussed in [23], an average heat transfer coefficient is better defined not as the surface average $\langle h \rangle$ of the local h, but rather as

$$h_{avg} = \frac{\langle q'' \rangle}{T_{bulk} - \langle T_{wall} \rangle} \tag{2}$$

in which the symbol $\langle\ \rangle$ denotes averaging operation on the active wall.

The Reynolds number (Re) is defined as:

$$\mathrm{Re} = \frac{UD_h}{\nu} \tag{3}$$

where D_h is the hydraulic diameter, conventionally assumed to be equal to twice the channel height (7.6 mm) as in a void plane channel of infinite width. U is the approach, or superficial, velocity, defined as the velocity that the fluid would have if the channel were void of any spacer; it is equal to the ratio between the flow rate and the passage area $U = Q/A_p$ (in the experiments $A_p = 9.12 \times 10^{-4}$ m^2). ν is the kinematic viscosity, which was assumed equal to 6.78×10^{-7} m^2/s in the cases investigated (water at ~43 °C).

For each configuration, the values of the flow rate investigated and the corresponding values of the Reynolds number are those given in Table 2. Flow rates were selected based on the experiments' limitations (the range of the flowmeters, measurement uncertainty at low flow rates, and pressure build up at the Plexiglas channel), and to enable comparison of results with several studies conducted in the literature by other researchers within the same range of values.

Table 2. Investigated values of the flow rate and corresponding Reynolds numbers (Re).

Q (L/min)	Re
1	205
1.5	307
2	410
2.5	512
3	615
3.5	717
4	820

2.2. Experimental Set-Up and Procedures

The experimental results were obtained in a previous study [35] using the setup shown in Figure 2. It is composed of two narrow (~4 mm height) hot and cold water channels separated by a 1 mm thick polycarbonate sheet. These channels are enclosed by an upper and a lower 2 cm thick transparent Plexiglass sheet. Test spacers were custom designed and fabricated using 3D printing with overall dimensions of 24 cm × 24 cm × 0.38 cm. The spacers were designed to fit and fill the hot channel while touching a thermochromic liquid crystals (TLC) sheet, which was made to adhere to the lower surface of the polycarbonate sheet. TLC sheets adopted had a color play from 35 to 40 °C. The hot and cold water temperatures were measured using RTD sensors located at the inlet and outlet of the two channels, while the flow rates were measured using flow sensors and rotameters. Julabo temperature-controlled heating and cooling circulators were used to control the hot and cold water temperatures at the channel inlets. Further details on experimental procedure and data processing can be found in [35]. The color distribution of the TLC at the upper surface of the hot channel was monitored by a high resolution camera, and images (1920 × 1080 pixels) were captured for different feed temperatures and flow rates (from 1 to 4 L/min), while the flow rate in the cold channel was kept at 8 L/min.

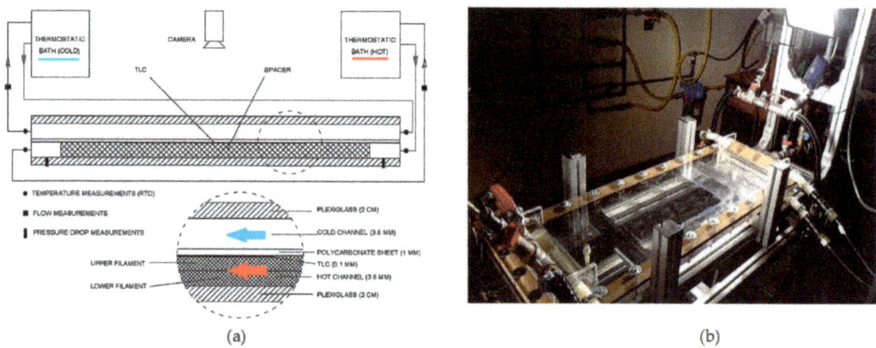

Figure 2. (**a**) Schematic representation and (**b**) picture of the experimental setup.

These images were processed by first converting them to the HSV (hue, saturation, value) format. It is known that the hue component of this format correlates with temperature but requires prior calibration. At the beginning of each spacer experiment, a TLC calibration test was preliminarily performed and a best fit polynomial curve was obtained for the temperature as a function of hue. During the calibration test both the upper and lower channels were fed by the same hot water source. In the real experiments, the cold water was set to 30 °C and the hot water was set to 43 °C, allowing a 13 °C difference between the two channels for heat transfer. Hot and cold water inlet temperatures were selected after several pre-tests to ensure that the temperature distributions on the TCL surface were clearly visible and fell within the color play range for all the flow rates.

It should be observed that in order to characterize the temperature polarization phenomenon, it is sufficient to reproduce the thermal field and convective heat transfer occurring in the spacer-filled channel only. This can be achieved by imposing at the active wall a realistic thermal boundary condition, representative of those occurring in MD modules, independent of whether there are or not an actual membrane and a vapor flux.

2.3. CFD Investigations

The problem was described by the continuity, momentum (Navier–Stokes), and energy equations for water at $T = 43$ °C and $p = 1$ bar, which were solved by the ANSYS-CFX® code (ANSYS, USA). In the simulations, the unit periodic cell approach was used [22,34,38], which simulates flow and heat exchange phenomena in periodic lattices under fully developed conditions (i.e., at sufficient distance from inlets and outlets). A source term, accounting for the large-scale temperature gradient, was implemented in the energy equation, and a body force per unit volume, accounting for the large-scale pressure gradient, in the momentum equations. In each run, this latter term was dynamically adjusted to obtain the required Re value (corresponding to one of those achieved in the experiments).

As reported in the literature, for geometries similar to those examined here [19,26], the fluid flow becomes unsteady for Re > 350. For this reason, steady-state laminar simulations were carried out for Re = 205 and 305, while, for 410 ≤ Re ≤ 820, the shear stress transport (SST) turbulence model was used, which is a blend between the k-ω model near the walls and the k-ε model in the outer region. As demonstrated in previous works [29], for the present transitional flows, ω-based models, which fully resolve the near-wall layer, are preferable to ε-based models, which make use of wall functions; among them, the SST model gives the most accurate predictions in terms of both distributions and average values of the heat transfer coefficient.

The computational domains (different according to the intrinsic angle) are shown in Figure 3, while details of the computational finite volume grid for case (c) are shown in Figure 4. Only four corner blocks were meshed with tetrahedra; the remaining volume was meshed with mapped hexahedra or with sweep mode in the regions adjacent to the cylindrical surfaces. In regard to the boundary conditions, periodicity was imposed at the walls perpendicular to the filaments; at the other walls a no slip condition was set. The fluid–filament interfaces and the bottom wall were assumed adiabatic, while at the top wall a mixed condition (Robin) was imposed by prescribing an external temperature of 30 °C and an external heat transfer coefficient equal to λ/δ, λ being the fluid's thermal conductivity and δ the channel thickness.

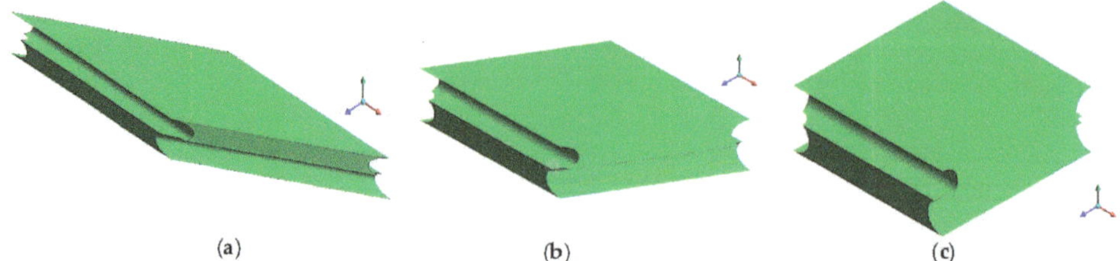

Figure 3. Computational domain (unit cell): (**a**) $\theta = 30°$; (**b**) $\theta = 60°$; (**c**) $\theta = 90°$.

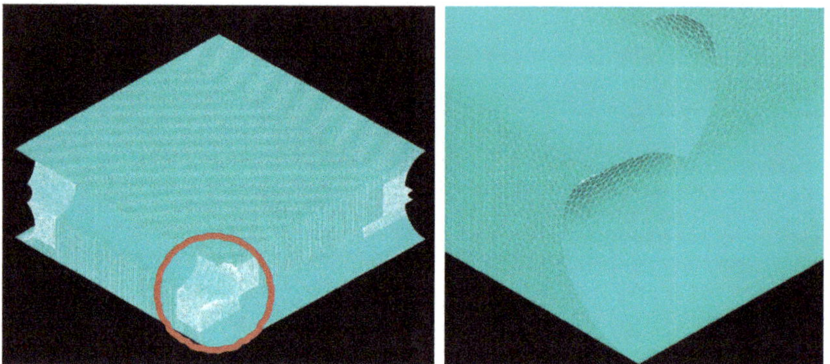

Figure 4. Mesh of a unit cell for case (**c**); one of the regions meshed with tetrahedra is highlighted.

3. Results and Discussion

3.1. CFD Prediction of Temperature and Heat Transfer Coefficient Distributions and Comparison with Experiments

For the sake of brevity, results are illustrated here in detail only for the two geometrical configurations in Figure 5. Configuration 1 is characterized by an intrinsic angle θ between filaments of 60° and a flow attack angle α of 30°, while configuration 2 is characterized by $\theta = 90°$ and $\alpha = 60°$. The solid profiles indicate the spacer filaments, which touch the upper, thermally active, wall. The results of the whole computational campaign (aiming at the determination of the overall dependence of $\langle h \rangle$ on Re for each geometrical configuration, to be used in process modelling applications) are reported in the final section.

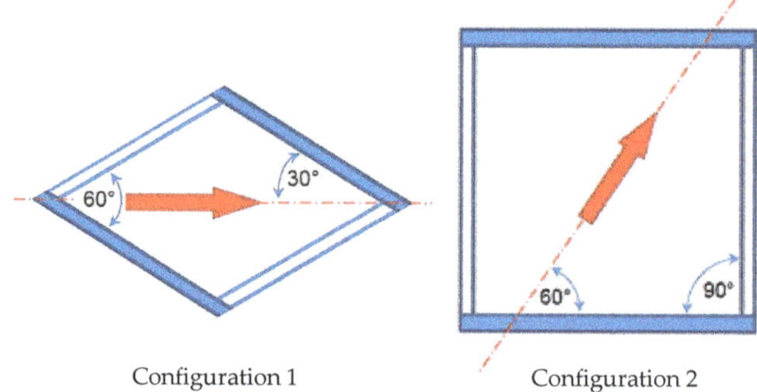

Figure 5. Sketch of the configurations for which experimental and computational distributions of the normalized heat transfer coefficient are compared. The intrinsic angle θ between the filaments and the flow attack angle α are reported.

For each configuration, three Reynolds numbers were considered: the lowest (205), an intermediate one (410), and the highest (820) of those reported in Table 2. Distributions of h, normalized by the mean value $\langle h \rangle$, are shown. Experimental distributions were obtained by post-processing the TLC images, as described in Section 2.2, and averaging the results over 3×3 unit cells to reduce dispersion.

3.1.1. Configuration 1

Results for configuration 1 are reported in Figure 6. For clarity purposes, the insets show the direction of the flow and the arrangement of the filaments.

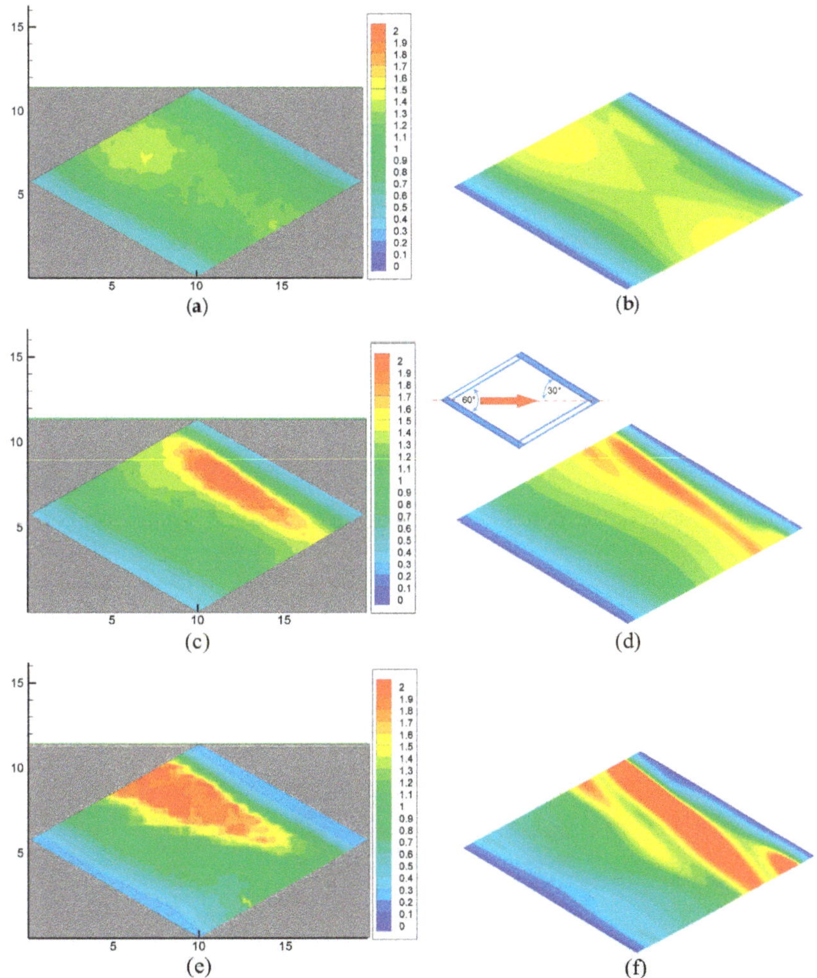

Figure 6. Configuration 1: distributions of the normalized heat transfer coefficient for increasing flow rates. (**a**,**b**) $Q = 1$ L/min. Re $\approx$ 205; (**c**,**d**) $Q = 2$ L/min. Re $\approx$ 410; (**e**,**f**) $Q = 4$ L/min. Re $\approx$ 820. Left column (**a**,**c**,**e**) experimental results; right column (**b**,**d**,**f**) computational fluid dynamics (CFD) predictions. Coordinates are in mm.

The comparison shows a fair overall agreement between simulations and experiments in the heat transfer coefficient distribution. At the lowest Re, the h distribution was roughly symmetric between the upstream (left) and downstream (right) halves of the cell, showing that inertial effects were negligible. As Re increased, a stripe of low *h* values grew shortly after the upstream upper filament, associated with a region of separated flow (as confirmed by an analysis of the CFD-predicted flow fields). At the same time, a band of high *h* values appeared shortly before the downstream upper filament, in correspondence with the reattachment of the separated shear layer. Some discrepancy existed between experiments and CFD results, which predicted stripes of high *h*, narrower, longer, and inclined with respect to the filaments. Note also that experimental results showed residual fluctuations, not eliminated by the spatial averaging performed (which only involved nine unit cells); CFD simulations, by their nature themselves, yielded time-averaged values of *h* (and of any other variable) and thus appeared smooth and regular.

3.1.2. Configuration 2

Results for configuration 2 are reported in Figure 7. As in the previous Figure 6, the inset shows the direction of the flow and the arrangement of the filaments.

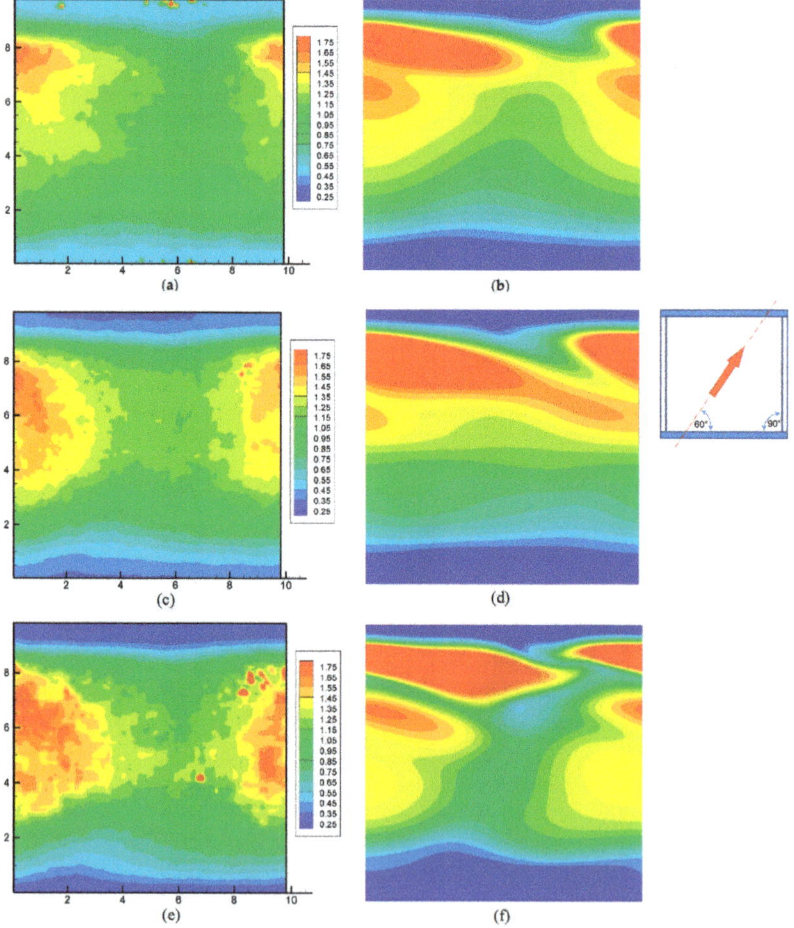

Figure 7. Configuration 2: distributions of the normalized heat transfer coefficient for increasing flow rates. (**a**,**b**) $Q = 1$ L/min. Re $\approx$ 205; (**c**,**d**) $Q = 2$ L/min. Re $\approx$ 410; (**e**,**f**) $Q = 4$ L/min. Re $\approx$ 820. Left column (**a**,**c**,**e**) experimental results; right column (**b**,**d**,**f**) CFD predictions. Coordinates are in mm.

In this case only a partial agreement between simulation and experiments could be registered. The numerical simulations predicted, both under laminar flow conditions (Figure 7b) and using the turbulence model (Figure 7d,f), a band where the most intense heat transfer occurs, located shortly before the downstream (upper) spacer filament, and a secondary band located shortly upstream of it (i.e., closer to the center of the unit cell). These features were not present in the experimental results, which showed a broad maximum of h only slightly upstream of the upper filament and centered on the opposite filament (the one that did not touch the active upper wall).

General levels and the overall distribution of h/h_{avg}, however, were satisfactorily reproduced. As in the previous configuration, experimental results exhibited a considerable amount of residual fluctuations, not completely damped by the 9-cell spatial averaging performed, while CFD results were smooth and regular.

3.2. CFD Prediction of Average Heat Transfer Coefficient and Comparison with Experiments

Figures 8 and 9 show the average heat transfer coefficient h_{avg} (Equation (2)) as a function of the Reynolds number for configurations 1 ($\theta = 60°$, $\alpha = 30°$) and 2 ($\theta = 90°$, $\alpha = 60°$), respectively.

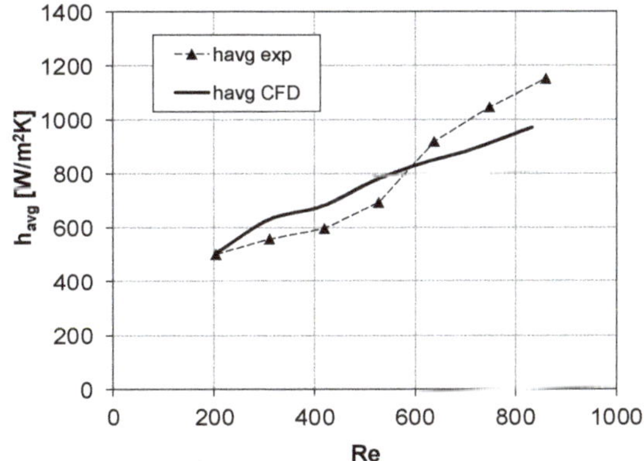

Figure 8. h_{avg} vs. Re for $\theta = 60°$, $\alpha = 30°$ (configuration 1). Symbols denote the experimental results, while CFD results are represented by the thick solid line.

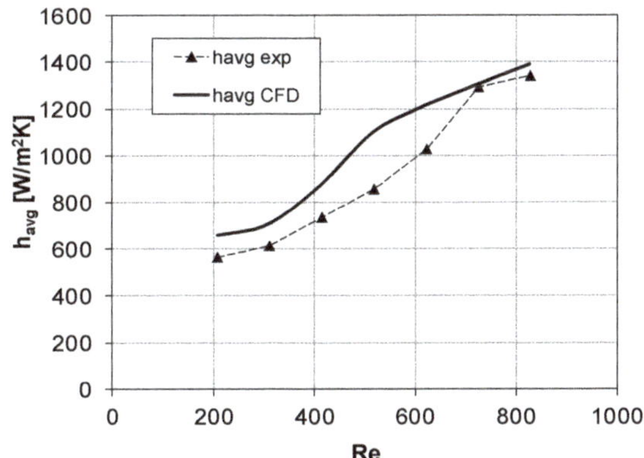

Figure 9. h_{avg} vs. Re for $\theta = 90°$, $\alpha = 60°$ (configuration 2). Symbols denote the experimental results, while CFD results are represented by the thick solid line.

A fair agreement between the experiments and the CFD simulations could be noticed. CFD results for configuration 1 (Figure 8) overpredicted h_{avg} by ~13% when Re was lower than ~600; at higher Re, a slight underprediction was registered (~14%). In Figure 9, results for configuration 2 showed that CFD overestimates h_{avg} by ~14%, with the maximum difference between experiments and simulations (~25% overprediction) observed at Re ≈ 500–600.

3.3. Summary Results for All Geometries Investigated

In Figures 10 and 11, the CFD results obtained for all the configurations reported in Table 1 are shown. In Figure 10, h_{avg} is plotted as a function of Reynolds number for the three configurations in which the flow direction bisects the intrinsic angle θ, i.e., $\theta = 30°$, $\alpha = 15°$; $\theta = 60°$, $\alpha = 30°$; and $\theta = 90°$, $\alpha = 45°$. Curves of h_{avg} against Re, for the configuration with $\theta = 90°$, are reported in Figure 11 considering flow attack angles values lower (graph a) or higher (graph b) than 45°.

From Figures 10 and 11, it is possible to identify the configuration with $\theta = 90°$, $\alpha = 60°$ as that providing the highest heat transfer coefficient.

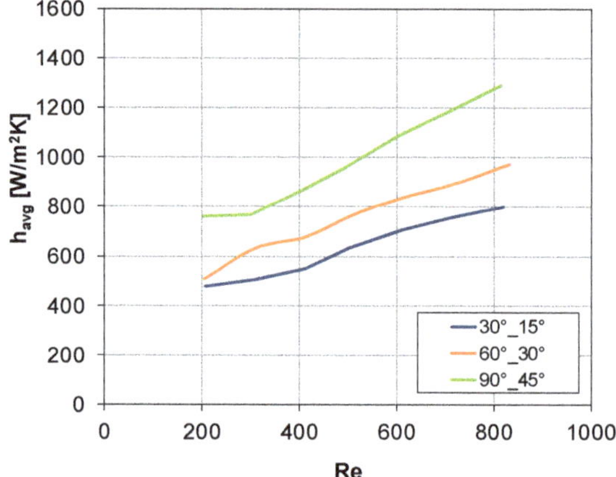

Figure 10. CFD results: h_{avg} vs. Re for $\theta = 30°$, $\alpha = 15°$; $\theta = 60°$, $\alpha = 30°$; and $\theta = 90°$, $\alpha = 45°$.

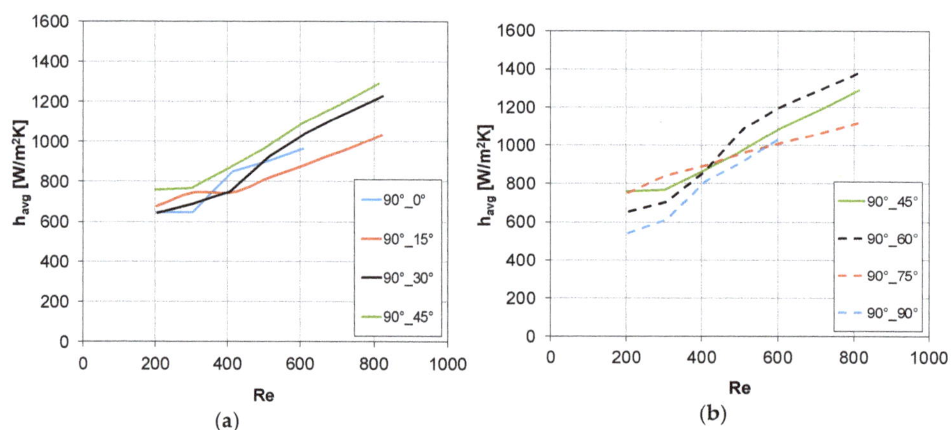

Figure 11. CFD results: h_{avg} vs. Re for $\theta = 90°$: (**a**) $\alpha \leq 45°$ and (**b**) $\alpha \geq 45°$.

4. Conclusions

Flow and heat transfer were predicted by computational fluid dynamics for channels provided with spacers consisting of two overlapped layers with a fixed pitch to height ratio (~2.63) and different values of the intrinsic angle between the filaments and of the flow attack angle. The Reynolds number ranged between 200 and 820.

In a previous work, experiments were carried out for the same configurations and Reynolds numbers by using thermochromic liquid crystals (TLC), which provide the distribution of the heat transfer coefficient h on the active wall.

For two selected cases, distributions of h obtained by CFD were compared with the experimental distributions, obtained by post-processing the real images of the TLC. From the comparison, a fair agreement concerning the shape of the h distributions was registered; the match was better at low Reynolds number, when no turbulence model was used, and worse at high Re, when simulations relied on the SST turbulence model.

For the same test cases, the average heat transfer coefficient was also fairly well predicted by CFD simulations, with discrepancies of the order of 10–20%, which can be regarded as only minor in view of the complex geometry and flow structure.

Finally, the results of the whole computational campaign were reported in terms of average heat transfer coefficient as a function of the Reynolds number and can be used in process modelling applications involving spacers with the same geometrical features as those investigated here.

In MD systems, geometrical configurations yielding the highest possible heat transfer coefficient h should be adopted in order to minimize the membrane surface (the largest component of plant cost). With reference to the geometries investigated in this work, for Reynolds numbers above 400–500, this optimal configuration seems to be characterized by an intrinsic angle θ of 90° and a flow attack angle α of 60°.

Author Contributions: Conceptualization, A.C., M.C., G.M., and M.A.; methodology, M.C. and M.L.C.; software, M.C. and M.L.C.; validation, M.A., N.T., and S.B.; formal analysis, M.C. and M.A.; investigation, M.L.C., N.T., and S.B.; resources, M.A. and A.C.; data curation, M.L.C. and M.C.; writing—original draft preparation, M.L.C.; writing—review and editing, A.C., M.C., and M.A.; visualization, M.L.C., N.T., and S.B.; supervision, A.C., M.C., and M.A.; project administration, M.A.; funding acquisition, M.A.

Funding: This project was supported by the NSTIP strategic technologies program in the Kingdom of Saudi Arabia, Project No. 11WAT1576-03.

Conflicts of Interest: The authors declare no conflict of interest.

References

1. Gude, V.G.; Nirmalakhandan, N. Sustainable desalination using solar energy. *Energy Convers. Manag.* **2010**, *51*, 2245–2251. [CrossRef]
2. Summers, E.K.; Arafat, H.A.; Lienhard, V.J.H. Energy efficiency comparison of single-stage membrane distillation (MD) desalination cycles in different configurations. *Desalination* **2012**, *290*, 54–66. [CrossRef]
3. Zuo, G.; Guan, G.; Wang, R. Numerical modeling and optimization of vacuum membrane distillation module for low-cost water production. *Desalination* **2014**, *339*, 1–9. [CrossRef]
4. Dong, G.; Kim, J.F.; Kim, J.H.; Drioli, E.; Lee, Y.M. Open-source predictive simulators for scale-up of direct contact membrane distillation modules for seawater desalination. *Desalination* **2017**, *402*, 72–87. [CrossRef]
5. Drioli, E.; Ali, A.; Macedonio, F. Membrane distillation: Recent developments and perspectives. *Desalination* **2015**, *356*, 56–84. [CrossRef]
6. González-Bravo, R.; Elsayed, N.A.; Ponce-Ortega, J.M.; Nápoles-Rivera, F.; El-Halwagi, M.M. Optimal design of thermal membrane distillation systems with heat integration with process plants. *Appl. Therm. Eng.* **2015**, *75*, 154–166. [CrossRef]
7. Subramani, A.; Jacangelo, J.G. Emerging desalination technologies for water treatment: A critical review. *Water Res.* **2015**, *75*, 164–187. [CrossRef]
8. Pangarkar, B.L.; Deshmukh, S.K.; Sapkal, V.S.; Sapkal, R.S. Review of membrane distillation process for water purification. *Desalin. Water Treat.* **2014**, *3994*, 1–23. [CrossRef]
9. Long, R.; Lai, X.; Liu, Z.; Liu, W. Direct contact membrane distillation system for waste heat recovery: Modelling and multi-objective optimization. *Energy* **2018**, *148*, 1060–1068. [CrossRef]

10. Wang, P.; Chung, T.S. Recent advances in membrane distillation processes: Membrane development, configuration design and application exploring. *J. Membr. Sci.* **2015**, *474*, 39–56. [CrossRef]
11. Fritzmann, C.; Lowenberg, J.; Wintgens, T.; Melin, T. State-of-the-art of reverse osmosis desalination. *Desalination* **2007**, *216*, 1–76. [CrossRef]
12. Ghaffour, N.; Lattemann, S.; Missimer, T.; Ng, K.C.; Sinha, S.; Amy, G. Renewable energy-driven innovative energy-efficient desalination technologies. *Appl. Energy* **2014**, *136*, 1155–1165. [CrossRef]
13. Martínez-Díez, L.; Vázquez-González, M. Temperature and concentration polarization in membrane distillation of aqueous salt solutions. *J. Membr. Sci.* **1999**, *156*, 265–273. [CrossRef]
14. Cao, Z. CFD simulations of net-type turbulence promoters in a narrow channel. *J. Membr. Sci.* **2001**, *185*, 157–176. [CrossRef]
15. Sousa, P.; Soares, A.; Monteiro, E.; Rouboa, A. A CFD study of the hydrodynamics in a desalination membrane filled with spacers. *Desalination* **2014**, *349*, 22–30. [CrossRef]
16. Amokrane, M.; Sadaoui, D.; Koutsou, C.; Karabelas, A.; Dudeck, M.; Karabelas, A. A study of flow field and concentration polarization evolution in membrane channels with two-dimensional spacers during water desalination. *J. Membr. Sci.* **2015**, *477*, 139–150. [CrossRef]
17. Koutsou, C.; Yiantsios, S.; Karabelas, A.; Karabelas, A. A numerical and experimental study of mass transfer in spacer-filled channels: Effects of spacer geometrical characteristics and Schmidt number. *J. Membr. Sci.* **2009**, *326*, 234–251. [CrossRef]
18. Taamneh, Y.; Bataineh, K. Improving the performance of direct contact membrane distillation utilizing spacer-filled channel. *Desalination* **2017**, *408*, 25–35. [CrossRef]
19. Gurreri, L.; Tamburini, A.; Cipollina, A.; Micale, G.; Ciofalo, M. Flow and mass transfer in spacer-filled channels for reverse electrodialysis: A CFD parametrical study. *J. Membr. Sci.* **2016**, *497*, 300–317. [CrossRef]
20. La Cerva, M.F.; Di Liberto, M.; Gurreri, L.; Tamburini, A.; Cipollina, A.; Micale, G.; Ciofalo, M. Coupling CFD with a one-dimensional model to predict the performance of reverse electrodialysis stacks. *J. Membr. Sci.* **2017**, *541*, 595–610. [CrossRef]
21. Saeed, A.; Vuthaluru, R.; Vuthaluru, H.B. Investigations into the effects of mass transport and flow dynamics of spacer filled membrane modules using CFD. *Chem. Eng. Res. Des.* **2015**, *93*, 79–99. [CrossRef]
22. Gurreri, L.; Tamburini, A.; Cipollina, A.; Micale, G.; Ciofalo, M. CFD simulation of mass transfer phenomena in spacer filled channels for reverse electrodialysis applications. *Chem. Eng. Trans.* **2013**, *32*, 1879–1884.
23. La Cerva, M.; Ciofalo, M.; Gurreri, L.; Tamburini, A.; Cipollina, A.; Micale, G. On some issues in the computational modelling of spacer-filled channels for membrane distillation. *Desalination* **2017**, *411*, 101–111. [CrossRef]
24. Al-sharif, S.; Albeirutty, M.; Cipollina, A.; Micale, G. Modelling flow and heat transfer in spacer-filled membrane distillation channels using open source CFD code. *Desalination* **2013**, *311*, 103–112. [CrossRef]
25. Koutsou, C.P.; Yiantsios, S.G.; Karabelas, A.J. Direct numerical simulation of flow in spacer-filled channels: Effect of spacer geometrical characteristics. *J. Membr. Sci.* **2007**, *291*, 53–69. [CrossRef]
26. Mojab, S.M.; Pollard, A.; Pharoah, J.G.; Beale, S.B.; Hanff, E.S. Unsteady laminar to turbulent flow in a spacer-filled channel. *Flow Turbul. Combust.* **2014**, *92*, 563–577. [CrossRef]
27. Ponzio, F.N.; Tamburini, A.; Cipollina, A.; Micale, G.; Ciofalo, M. Experimental and computational investigation of heat transfer in channels filled by woven spacers. *Int. J. Heat Mass Transf.* **2017**, *104*, 163–177. [CrossRef]
28. Shakaib, M.; Hasani, S.M.F.; Ahmed, I.; Yunus, R.M. A CFD study on the effect of spacer orientation on temperature polarization in membrane distillation modules. *Desalination* **2012**, *284*, 332–340. [CrossRef]
29. Ciofalo, M.; La Cerva, M.; Di Liberto, M. Selection of Turbulence Models for Low-Reynolds Number Turbulent Flow in Spacer-Filled Channels. In Proceedings of the 36th UIT Heat Transfer Conference, Catania, Italy, 25–27 June 2018.
30. Chang, H.; Hsu, J.A.; Chang, C.L.; Ho, C.D.; Cheng, T.W. Simulation study of transfer characteristics for spacer-filled membrane distillation desalination modules. *Appl. Energy* **2017**, *185*, 2045–2057. [CrossRef]
31. Katsandri, A. A theoretical analysis of a spacer filled flat plate membrane distillation modules using CFD: Part I: Velocity and shear stress analysis. *Desalination* **2017**, *408*, 145–165. [CrossRef]
32. Tamburini, A.; Pitò, P.; Cipollina, A.; Micale, G.; Ciofalo, M. A Thermochromic Liquid Crystals Image Analysis technique to investigate temperature polarization in spacer-filled channels for Membrane Distillation. *J. Membr. Sci.* **2013**, *447*, 260–273. [CrossRef]

33. Tamburini, A.; Cipollina, A.; Al-Sharif, S.; Albeirutty, M.; Gurreri, L.; Micale, G.; Ciofalo, M. Assessment of temperature polarization in membrane distillation channels by liquid crystal thermography. *Desalin. Water Treat.* **2014**, *55*, 37–41. [CrossRef]
34. Tamburini, A.; Renda, M.; Cipollina, A.; Micale, G.; Ciofalo, M. Investigation of heat transfer in spacer-filled channels by experiments and direct numerical simulations. *Int. J. Heat Mass Transf.* **2016**, *93*, 1190–1205. [CrossRef]
35. Albeirutty, M.; Turkmen, N.; Al-Sharif, S.; Bouguecha, S.; Malik, A.; Faruki, O.; Cipollina, A.; Ciofalo, M.; Micale, G. An experimental study for the characterization of fluid dynamics and heat transport within the spacer-filled channels of membrane distillation modules. *Desalination* **2018**, *430*, 136–146. [CrossRef]
36. Phattaranawik, J.; Jiraratananon, R.; Fane, A. Heat transport and membrane distillation coefficients in direct contact membrane distillation. *J. Membr. Sci.* **2003**, *212*, 177–193. [CrossRef]
37. Phattaranawik, J.; Jiraratananon, R.; Fane, A.G.; Halim, C. Mass flux enhancement using spacer filled channels in direct contact membrane distillation. *J. Membr. Sci.* **2001**, *187*, 193–201. [CrossRef]
38. Gurreri, L.; Tamburini, A.; Cipollina, A.; Micale, G.; Ciofalo, M. CFD prediction of concentration polarization phenomena in spacer-filled channels for reverse electrodialysis. *J. Membr. Sci.* **2014**, *468*, 133–148. [CrossRef]

Article

Characterization and Assessment of a Novel Plate and Frame MD Module for Single Pass Wastewater Concentration–FEED Gap Air Gap Membrane Distillation

Rebecca Schwantes [1,2,3,*], Jakob Seger [4], Lorenz Bauer [1], Daniel Winter [2], Tobias Hogen [4], Joachim Koschikowski [2] and Sven-Uwe Geißen [4]

1 SolarSpring GmbH, 79114 Freiburg, Germany
2 Fraunhofer Institute for Solar Energy Systems, 79110 Freiburg, Germany
3 Institute of Power Engineering, Technische Universität Dresden, 01062 Dresden, Germany
4 Department of Environmental Technology, Technische Universität Berlin, 10623 Berlin, Germany
* Correspondence: rebecca.schwantes@solarspring.de

Received: 15 August 2019; Accepted: 2 September 2019; Published: 6 September 2019

Abstract: Membrane distillation (MD) is an up and coming technology for concentration and separation on the verge of reaching commercialization. One of the remaining boundaries is the lack of available full-scale MD modules and systems suitable to meet the requirements of potential industrial applications. In this work a new type of feed gap air gap MD (FGAGMD) plate and frame module is introduced, designed and characterized with tap water and NaCl–H_2O solution. The main feature of the new channel configuration is the separation of the heating and cooling channel from the feed channel, enabling a very high recovery ratio in a single pass. Key performance indicators (KPIs) such as flux, gained output ratio (GOR), recovery ratio and thermal efficiency are used to analyze the performance of the novel module concept within this work. A recovery rate of 93% was reached with tap water and between 32–53% with salt solutions ranging between 117 and 214 g NaCl/kg solution with this particular prototype module. Other than recovery ratio, the KPIs of the FGAGMD are similar to those of an air gap membrane distillation (AGMD) channel configuration. From the experimental results, furthermore, a new MD KPI was defined as the ratio of heating and cooling flow to feed flow. This R_F ratio can be used for optimization of the module design and efficiency.

Keywords: distillation; high recovery rate; brine concentration; zero liquid discharge; membrane distillation module; wastewater concentration; resource recovery

1. Introduction

Membrane distillation (MD) has drawn increasing interest in the last 10 years from both academia and applicational parties. Beginning with a wide range of basic testing of various solutions [1–6] and the establishment of thermodynamic process understanding and modelling [7–9], the state of research has gradually shifted towards a more applied focus, bringing forth more advances in fields of bench and pilot scale trials, membrane testing [10,11] and investigations on membrane hydrophobicity as well as the regeneration of such [12–19]. An increase in modelling, targeted on module and system or even hybrid system scenarios [20,21] together with tecno-economic calculations for industrial applications [22,23] can also be observed. Overall, a shift of perception is gradually taking place in which MD is no longer seen as strongly as a possible substitute for seawater reverse osmosis (SWRO), but its potential as a process step in brine or wastewater concentration and recycling is being unfolded; this is especially valid in cases where reverse osmosis (RO) reaches its functional boundaries due to high osmotic pressure differences. MD is also being adopted in other applications e.g., in the pharmaceutical industry or food industry because it can be operated at ambient pressure and with low temperatures [24–26].

Despite these increasingly practical advances, MD is not fully commercialized yet and some of the reasons repeatedly given are the lack of full-scale MD modules, high specific thermal energy consumption, low recovery rate and a limited variety of available specialized membranes [27,28]. For example, the regeneration of membrane hydrophobicity in situ and the avoidance of membrane wetting altogether has not yet brought forth a universal strategy for commercial implementation in flat sheet MD modules on a large scale, especially when dealing with high salinity feed solutions—although increasingly targeted by research in the last few years [29,30]. However, promising results have been achieved for specific applications using brines between 1–19% and using porous fluorosiloxane-coated polypropylene hollow fibers [31,32].

In addition, technology requirements in industrial applications can be different than those present during the technology development conducted at a time when solar-powered seawater desalination was still the main goal of many MD researchers and developers. With this in mind, and based on the previous experience, new adaptations to MD modules and systems are necessary in conjunction with piloting in the relevant industry.

Such an opportunity was created by the HighCon project, placed within the funding program WavE–"Future-oriented Technologies and Concepts to Increase Water Availability by Water Reuse and Desalination" [30]. The program was launched by the German Federal Ministry of Education and Research in order to facilitate the adoption of more comprehensive and ecofriendly wastewater treatment and internal reuse of resources in the industry. The project is coordinated by the Technische Universität Berlin, Germany. Within HighCon, the goal is to recover resources from concentrates generated by recycling of industrial wastewater at two different pilot sites. Membrane distillation was represented by the company SolarSpring GmbH and the Fraunhofer Institute for Solar Energy Systems within the consortium. During the project, a zero liquid discharge process chain was developed and piloted, based on a chain of membrane technologies and a humidification-dehumidification (HDH) crystallization step as the final stage.

Due to the necessity to concentrate the pretreated saline wastewater to a near saturation concentration level in the MD system for subsequent crystallization, a redesign of the MD module and system was carried out and then put into practice in a full-scale prototype. The three main goals to be achieved with the new concept are:

(I) Implementing a plate and frame module design with replaceable inner parts;
(II) Minimization of the number of components in contact with the highly corrosive feed solution;
(III) Enabling of a single pass recovery rate high enough to operate the MD stage as a one- step process.

These design criteria apply to almost any industrial process in which highly corrosive or chemically aggressive or toxic compounds are present in the feed solution and the concentration increase should be high. So far, in the existing MD channel configurations that are based on sensible heat recovery and using flat sheet membranes, the unification of points (I)–(III) were not possible on a full scale and is considered a drawback of the technology. Also, the replacement of inner parts is not possible per se in any module type in which a resin potting is used for sealing of the material layers. Low module recovery rates also mean that to achieve a high increase in concentration, the feed must be circulated in a batch mode or processed by a serial cascade of modules [33]. This requires commercially available heat exchangers and pumps that are chemically resistant to the feed solution to be treated. In this work a new module concept is presented under the name of feed gap air gap membrane distillation (FGAGMD) which incorporates the solution to requirements (I)–(III). Fundamentally, it is a derivation of air gap membrane distillation (AGMD) [34,35], however the feed and the permeate solution are separated from the heating and cooling solution by addition of an extra channel. Precise details on the configuration are provided in Sections 2.1 and 2.2. Furthermore, the new FGAGMD module was constructed and characterized within the HighCon project. A comprehensive description on the pilot system set-up and experimental configuration is given in Section 2.3. Full characterization data for tap water and artificial NaCl–H_2O solution with a salinity up to 214 g NaCl/kg is provided in this work and discussed in the results Section 3.

Regarding the topic of specific thermal energy consumption in MD as such, there is still a lack of consensus in publications to provide the key performance indicators necessary to properly evaluate the entire performance of the process. As explained in [22], experiments carried out in small-scale bench testing, are often presented using flux values without any mention of energy consumption [33], thus giving a misleading impression on how future process optimization needs to be carried out. Simply targeting a high gained output ratio (GOR) in MD module design will, however, not satisfy the future market demands, as it is by nature of the process always a trade-off with transmembrane flux, the latter reducing with increasing GOR [36–38]. Thus, developments in MD module and process design should be carried out keeping both in mind as described by [22,38]. The question if to maximize GOR or flux should be evaluated carefully on a case to case basis when it comes to the implementation in the industry. Within this work flux, GOR, recovery ratio and thermal efficiency are used to gain a comprehensive understanding of the module performance.

2. Materials and Methods

2.1. Membrane Distillation (MD) Process and the Evolution of Channel Configurations

Membrane distillation is a thermal driven membrane process in which a hydrophobic, microporous membrane constitutes the vapour space. Only vapours can pass through the membranes, non-volatile substances are rejected. The driving force of the process is the effective vapour pressure difference across the membrane most commonly provided by a temperature difference. Mass transfer in MD is well investigated [9,39] and can be described as a combination of molecular and Knudsen diffusion transport mechanisms. Heat transfer in MD is characterized by a combination of latent heat and conductive heat transfer and is sensitive to many parameters such as physical membrane properties, spacer selection and operational parameters [40]. Since the early days of MD [41], the possible MD channel configurations and process designs have evolved in several different directions [42–45]. One type of MD system, commercially established by the company memsys [46], is a combined thermal and vacuum driven process named VMEMD [47]. Essentially, it represents the process commonly known as multi-effect distillation (MED) equivalent of MD. In this variant the latent heat is recovered and passed on from stage to stage. When talking about plate and frame modules in this work however, the focus will be on the MD process variants in which sensible heat is recovered to increase the efficiency, thus representing a multi-stage flash (MSF) type of heat recovery principle if thought of in terms of state-of-the-art evaporation technology.

In the following, new derivations of MD channel configurations are introduced. An overview of the commonly known configurations direct contact membrane distillation (DCMD) and AGMD [48], as well as the new configurations feed gap membrane distillation (FGMD) and FGAGMD is provided in Figure 1.

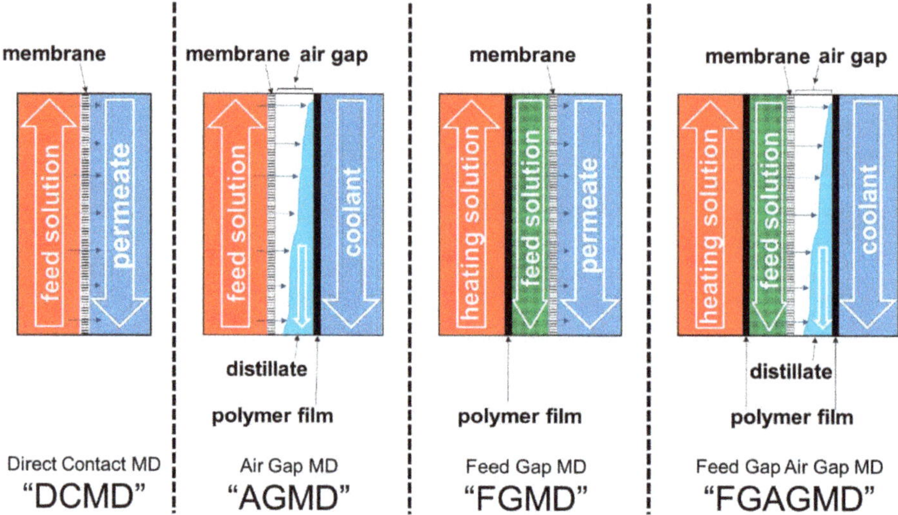

Figure 1. Basic channel configurations direct contact membrane distillation (DCMD) and air gap membrane distillation (AGMD) and derived "Feed gap" variant feed gap membrane distillation (FGMD) and feed gap air gap MD (FGAGMD).

The DCMD is the simplest MD configuration in which a hot feed solution flows on one side and the cooled permeate flows on the other side of the microporous membrane in counter current. Due to the temperature-induced vapour pressure gradient, vapour is transported through the membrane pores from the hot to the cold side of the membrane which then condenses on entering the permeate stream. An advantage of this configuration is the high flux due to the low thermal barrier of only the membrane material enabling high flux and thermal efficiency at low salinities [36]. However, the fraction of conductive heat passed through the membrane is also comparably higher leading to a potentially lower thermal efficiency in modules with significant heat recovery and especially at higher salinities above 100 g NaCl/kg. There is enough evidence in research to acknowledge that the quality of distillate can, furthermore, be compromised by so called wetting phenomena and that these phenomena are enhanced when there is liquid on both sides of the membrane [14,49]. In DCMD this is unavoidable per se und must, therefore, be considered a disadvantage. Another system-related drawback is the requirement of an additional heat exchanger in order to recover heat from the condenser outlet for pre-heating of the feed. In summary, DCMD is very popular in academia for its simplicity and high flux, but has limitations when it comes to use in the industry and is not as commonly applied on a full scale as e.g., AGMD or PGMD [50–54].

The configuration second from the left in Figure 1 is AGMD. In this variant, a thin polymer film separates the distillate channel from the coolant. This also enables the direct internal recovery of heat inside the module, since the feed solution can thereby be used as coolant without. The additional air gap in AGMD also provides a higher thermal insulation between the channels. This reduces the overall heat transfer, thus lowering the flux compared to DCMD but it also reduces the fraction of conductive heat transfer which is not used for evaporation. A comprehensive comparison of the two variants DCMD and AGMD is given by [48]. At higher salinities, DCMD can be more sensitive to the impact of vapour pressure reduction [22] unless thicker membranes are used in which case the thicker membrane substitutes the thermal insulation properties of the air gap [55]. The AG channel variant also gives the possibility of applying a vacuum (v-AGMD) to reduce molecular diffusion resistances or to use low-pressure air sparging to ensure a complete drainage of the distillate, the positive effects of which on wetting are reported in [49].

One feature of both DCMD and AGMD are limitations regarding the recovery ratio of distillate in a single pass through the module. Since the thermal energy for evaporation is carried into the module via the feed stream on the evaporator side, the permeate output is limited by the sensible heat contained in that stream. The maximum theoretical recovery ratio in a single pass can be calculated by dividing the latent heat of evaporation at the mean process temperature by the sensible heat supplied be the heating channel. As presented in detail by [38], only approximately 1–8% of the evaporator inlet mass flow can be extracted as permeate in the aforementioned channel configurations. For example, to achieve a concentration increase from 8% *w/w* to ~20% *w/w* of NaCl H_2O solution, a recovery ratio of 60% would be necessary in order to reach this goal in a single pass. In order to decouple the thermal energy supply from the feed stream and overcome the limitation of recovery ratio, one possibility is an alternative operation of an AGMD channel set-up. In this so- called FGMD configuration, the feed solution is circulated through the previous air gap and the heating solution is introduced into the former coolant channel which is separated from the feed solution by the polymer film. On the cold side of the membrane, the permeate is now circulated as coolant and the vapours condense in that coolant channel in analogy to the DCMD set-up as shown in Figure 1. Even though a separation of energy supply and feed supply is achieved thereby, the FGMD configuration still suffers from the same drawbacks as the DCMD arrangement regarding module internal heat recovery, sensitivity towards vapour pressure reduction and the need for an extra heat exchanger to recover the heat from the permeate channel outlet to preheat the feed. Thus, in order to fulfill all three goals listed in the introduction a further modification is needed. FGAGMD describes a variant in which not only the feed solution is separated from the heating stream, but the cooling stream is also separated from the permeate channel. Such a channel arrangement has previously been discussed in conjunction with MD process integration into heat exchangers by [56,57], but no full scale modules are available to the knowledge of the authors. Due to the addition of the extra second polymer film an additional conductive heat transfer resistance is added to the thermodynamic resistance chain. Also, it must be noted that an amount of sensible heat is transported out of the system together with the heated feed solution in the FGMD and FGAGMD channel configurations and is dependent on the feed solution flow rate, physical properties and temperature. This must be accounted for in the design of the overall system design of a potential FGMD or FGAGMD module and system, if this exiting heat is to be recovered. Due to the higher suitability of an air gap configuration for the concentration of solutions with a high vapour pressure reduction compared to pure H_2O and the advantages regarding the recovery of heat inside the module, a FGAGMD channel configuration was chosen for the pilot system and developed within the HighCon project. In addition, goals II and III can only be achieved with a separation of heat supply and feed supply to the module(s). Further advantages of the configuration are worth mentioning. The heating and cooling loop can be operated with a single load of water and easily connected directly to available heating and cooling sources without interfering with the feed stream.

2.2. Module Design

The goals of MD module design can be outlined as packaging the process into a device that facilitates a uniform flow distribution in the channels, and reduces energy demand, pressure losses and polarization effects. Cleaning, maintenance and replacement of components should be possible at a reasonable cost. Packing density should be high. All materials used must be mechanically, chemically and thermally adequate for the targeted application of the module [28].

A plate and frame FGAGMD module was designed, with an effective channel length of 5.76 m and channel height of 0.72 m. The channel length in an MD module with sensible heat recovery determines the temperature difference which will be established across the membrane. In opposition to a spiral wound module, the possible channel lengths in a plate and frame type are limited by the length of the individual plates that constitute the channel. The effective channel length can only be a multiple of the single plate length by a serial connection of plates. The single plates are hydraulically connected to each other via switch plates in order to achieve a 180° angle change in flow direction at the end of each plate. This way thermodynamic adaptations to specific customer needs can be implemented easily by adding or subtracting a certain number and arrangement of plates. Figure 2 provides a detailed overview of how the FGAGMD channel configuration was implemented in the prototype module. In order to achieve a minimal heat loss to the ambient, the channels on the respective outer sides of the module were designed to be cold channels. Thus, the only way to package the channel configuration, is to locate the hot channel in the middle of the individual channel set-up. In the schematic in Figure 2, the channels are indicated by the letters H (heating channel), C (cooling channel), D (distillate) and F (feed). The membrane is between each feed (F) and distillate channel (D) and the heating and feed as well as the cooling and distillate channels are separated by a thin polypropylene film. The numbers in the channels and plates of the schematic give reference to the respective material properties in Table 1. Distillate permeation from the feed channels (F) to the distillate channels (D) occurs due to driving force temperature difference established between the heating channel (H) and the cooling channel (C). The basic thermodynamic principles of MD are well published [9,40,58,59] and will not be repeated in detail here. The evaluation of the new module concept will be carried out with use of common key performance indicators presented in Section 2.4. However, the differences of the new FGAGMD to a standard AGMD configuration should be briefly explained. As presented in Figure 3, in FGAGMD an additional conductive resistance is added through the extra heating channel and polymer film. In AGMD the heating channel is equivalent to the feed channel. The second decisive difference is the heat carried in and out of the system with the feed channel. Depending on the operation parameters, sensible heat leaves the system with the outlet of the feed solution. One method to recover this heat is shown in the experimental set-up of the pilot system explained in Section 2.3.

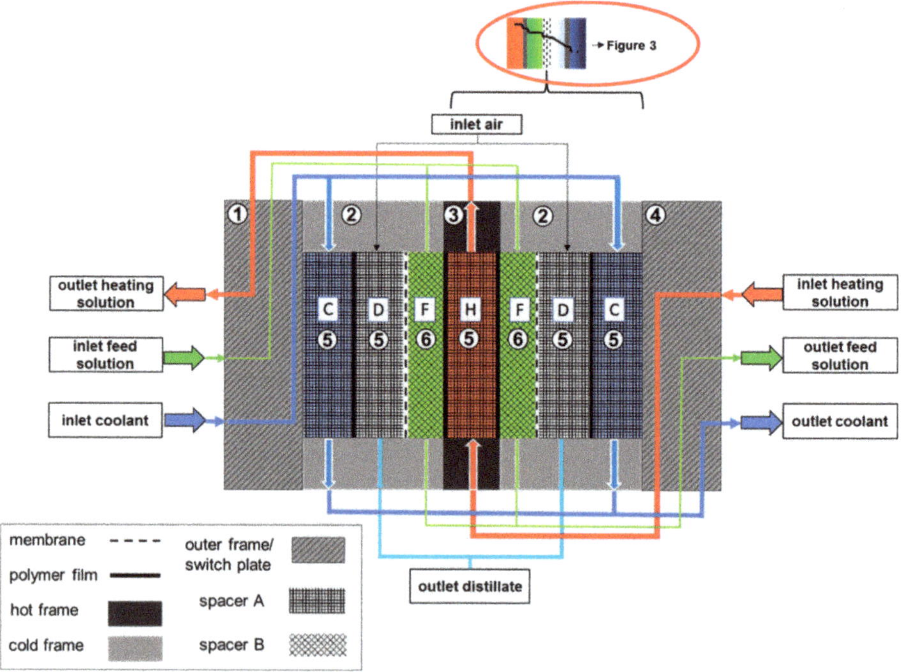

Figure 2. Detailed schematic overview of the FGAGMD channel configuration.

Table 1. Material properties of sub-components used inside the module.

No	Material	Thickness	Polymer	Approx. Porosity (%)	Nominal Pore Diameter (µm)
1 + 4	Outer- and switch plate	30 mm	Polypropylene (PP)	-	-
2 + 3	Hot and cold channels	4 mm/2 mm	PP	-	-
5	Spacer A	2 mm	High density Polyethylene (HDPE)	80	-
6	Spacer B	1 mm	PP	80	-
-	Membrane (+ backing)	76 (+280) µm	Polytetrafluorethylene (PTFE), (PP)	80 (50)	0.2
-	Polymer film	100 µm	PP	-	-

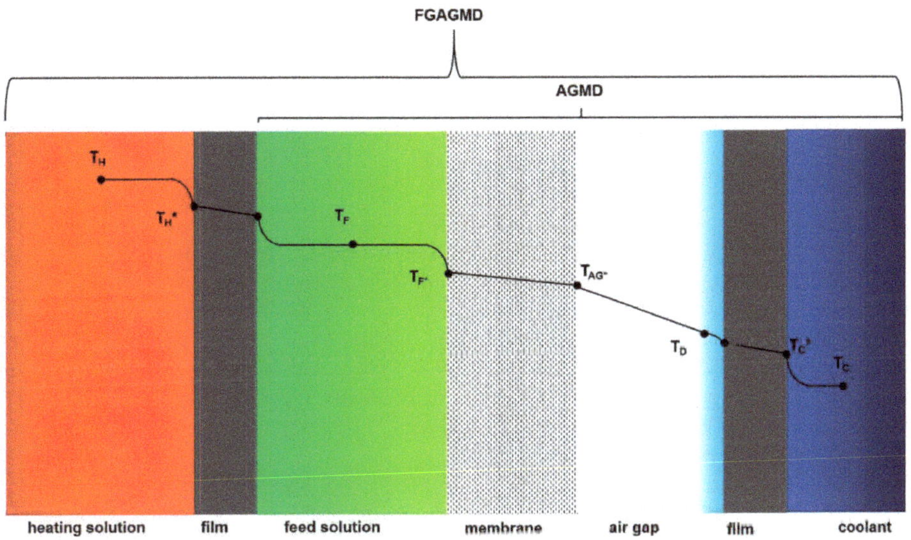

Figure 3. Temperature progression along the cross-section of the channels.

A closer look at the temperature progression from the hot channel to the right-side cold channel is shown in Figure 3. As indicated in Table 1, the thickness of the heating channel is twice that of the cooling channel since the capacities of the total heating and cooling flow should be as similar as possible for the sake of process efficiency as shown by [38]. The schema also shows how the FGAGMD set up adds to the loss of effective temperature driving force difference at the membrane interface. The loss is expressed by the difference between T_{H*} and T_{F*} and can be assigned directly to the conductive heat loss through the additional film between the feed and the heating solution as well as the additional thermal boundary layers in the feed solution. An assembled module is shown in Figure 4. Including the outer reinforcements, the dimensions of the module are 1 m × 2 m and the inner channel dimensions of a single plate are 0.72 m × 1.44 m. The membrane area of the individual modules is ~8 m^2. This module's shell was designed to hold a much higher membrane surface than implemented in this pilot system of up to 50 m^2. By parallelizing multiple channels, the packing density can thus be improved significantly.

Figure 4. Plate and frame FGMD module.

2.3. Experimental Set-Up

The pilot system consists of an FGAGMD module pair and a hydraulic system and heating on board. The modules are identical twins. It is important to note that only system components were considered, that are freely available in the market and declared as suitable for the application by the manufacturer. The heating and cooling loop is operated in parallel, however, the direction of feed flow is serial. As visible in Figure 5, the feed flow in module 1 is in parallel to the cooling channel whereas in module 2 it is parallel to the heating channel. This enables the recovery of the sensible heat leaving MD 1 in the second module MD 2. For a better understanding of the pilot set-up the flows and components shown in Figure 5 are described as follows:

2.3.1. Feed Loop

Using feed pump P_{f}, feed solution with the properties recorded by sensors T_{fi1} (temperature feed in MD 1), p_{fi1} (pressure feed in MD 1) and C_{fi1} (conductivity feed in MD 1) and a volume flow set by flow meter F_{fi1} is pumped from the feed tank into the first module (MD 1). After entering, the temperature of the feed solution is increased by the heating solution in a counter current manner through the thin polymer film. Thus, a thermal driving force is established between the feed solution and the cooling channel, which results in a vapour flux passing from the feed solution to the adjacent distillate channel. The remainder of the feed solution (flow F_{fo1}), leaves MD 1 at temperature T_{fo1} and at a conductivity of C_{fo1} and enters MD 2 at temperature T_{fi2} which is lower than T_{fo1} due to small losses to the ambient. Similarly, in MD 2 a vapour flux to the distillate channel is established via the temperature-induced vapour pressure difference. However, in MD 2 the feed channel flow is co-current to the heating channel. It exits MD 2 at temperature T_{fo2} and conductivity C_{fo2} and flows either back into the feed solution tank or leaves the system depending on the position of the control valve V_{f}. It is notable that the feed stream is not actively heated or cooled by any sources other than the MD modules. This means, that besides the feed tank, feed pump, piping and sensors, no other components require resistance towards highly concentrated salt solutions (or potentially other chemicals).

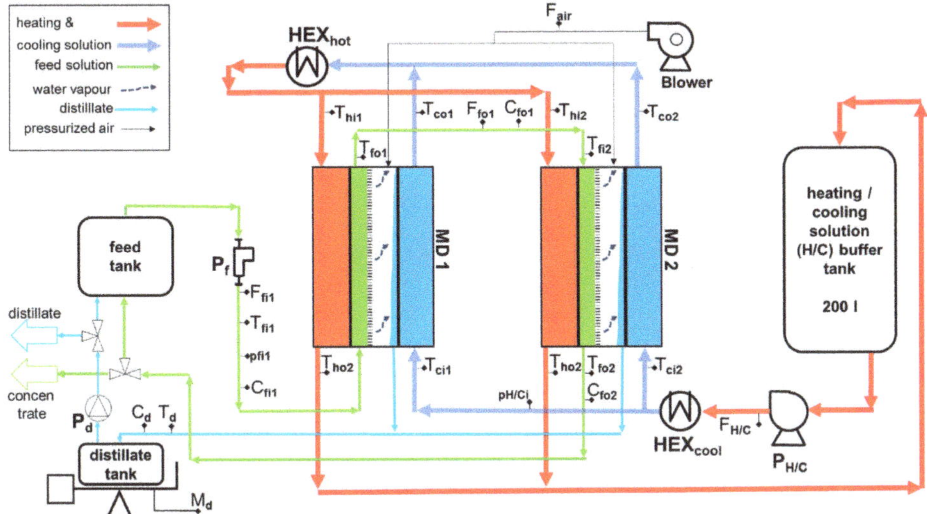

Figure 5. Schematic of the FGAGMD pilot system.

2.3.2. Heating and Cooling Loop

The heating and cooling (H/C) solution is pumped from the H/C solution buffer tank with pump P_{HC} controlled by flow meter F_{HC}. No specific materials need be used here, since the heat transfer fluid consists of softened tap water. In case of a leakage, a system shut-down would be triggered to protect the non-salt water resistant components in this loop. Similarly, pressure sensor p_{HCi} has a safety function. It protects the modules from overpressure but also provides a recorded value for the monitoring of pressure during dynamic operations. Before entering both modules in parallel, the H/C solution is cooled to a set temperature via heat exchanger HEX_{cool}. T_{ci1} and T_{ci2} record the respective cooling channel inlet temperatures of MD 1 and 2 of which T_{ci1} is the control sensor. The cooling channels are separated from the distillate channels by a thin polymer film. The coolant gains temperature due to the recovery of heat from the combination of conductive heat transfer and the latent heat of the distillate condensing on the film. At the outlet, the temperatures are recorded by T_{co1} and T_{co2}. The required external heat is supplied by HEX_{hot} after which the cooling solution becomes the heating solution by definition. In the pilot plant, the thermal energy is supplied by a heating rod and controlled to a fixed temperature monitored by sensor T_{hi1}. After the heating and cooling solution has entered the module, its temperature reduces constantly along the channel due to the heat transfer through the hot side polymer film in order to provide the process heat to the feed. It exits the heating channel at temperatures T_{ho1} and T_{ho2}, before flowing back to the H/C buffer tank.

2.3.3. Distillate

The diffusing water vapour enters the distillate channel "D" through the membrane and condenses on the surface of the condenser channel. The distillate drains either through gravity or is actively drained by low pressurized air supplied by a blower [49]. The volume flow rate of the air is recorded by sensor F_{air}. The distillate leaves both modules at a mixed temperature T_d and a conductivity C_d and is collected in the distillate tank. The tank mass is continuously recorded by the balance M_d. At a set mass value, pump P_d is triggered and the distillate is pumped either back to the feed tank, for operation continuous mode or discharged to the CIP tank to be used for the periodic cleaning of the module alongside other purposes. The CIP system is not shown in Figure 5 as its details are not relevant to this work, other than the fact that flushing of the module is possible in a fully automatic manner. Figure 6 shows an image of the entire pilot system with indication of the main components.

Permanent logging of all sensor values took place during operation to allow a high resolution of data points and, therefore, an adequate accuracy in data analysis. The logging frequency was set to 30 s per datapoint. Table 2 provides a complete overview of sensors and actors used in the system alongside their accuracy (if applicable) as indicated by the manufacturer.

Table 2. Sensor and actor list of the pilot system.

Component	Producer	Type	Accuracy	Name in Schematic
Conductivity meters	Jumo, Fulda, Germany	CTI500 24VDC	≤0.5% of measuring range (0–500 mS/cm)	C_{fi1}, C_{fo2}
Conductivity meter distillate	Jumo, Fulda, Germany	BlackLine CR-EC	≤2% of measuring range (0–5000 µS/cm)	C_d
Volume flow H/C	Krohne, Duisburg, Germany	Optiflux 4300C	0.5% of measuring value	$F_{H/C}$
Volume flow Feed	MIB GmbH, Breisach am Rhein, Germany	Flowmax 42i	±2% of measured value ± 3 mm/s	
Temperature sensors	TC direct, Mönchengladbach, Germany	Pt100 Klasse A	±(0.15 + 0.002 × t)	all Temperatures
Feed pump	KNF Neuberger GmbH, Freiburg, Germany	PML14169-NF 300	-	P_f
Heating and Cooling pump	Dunkermotoren, 79848 Bonndorf im Schwarzwald, Germany	BG 65X50 SI	-	
Pressure sensors	Jumo, Fulda, Germany	Midas, C18 SW -1,6	1.6% of measuring value	$p_{H/Ci}$, p_{fi1}
PLC	Advantech Europe BV, Hilden, Germany	2271G-E1-C20170517	-	-
Pressurized air pump	KNF, Neuberger GmbH, Freiburg, Germany	KNF N828 KNE	-	Blower
Balance distillate	Soehnle Industrial Solutions GmbH, Backnang, Germany	Table balance	1 g for 0–32 kg	M_d
Volume flow air	First Sensor AG, Berlin, Germany	WTA	±2% of reading +0.25% of measuring value	F_{air}
Heating and cooling pump	Harton Anlagentechnik GmbH, Alsdorf, Germany	DC 40/10BL	-	$P_{H/Ci}$

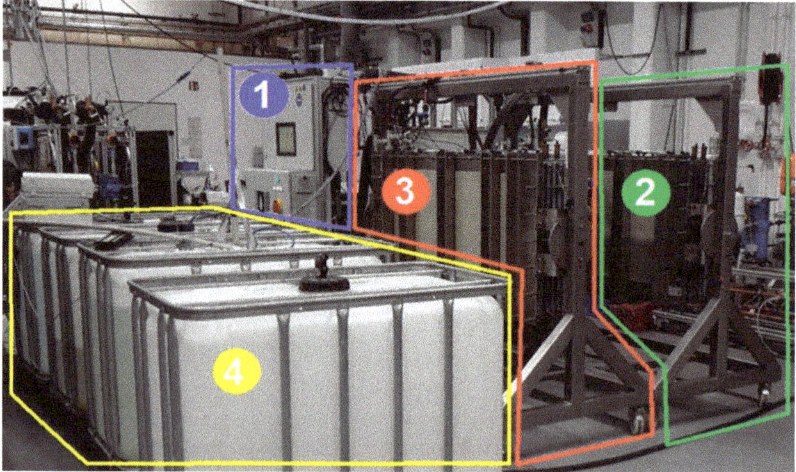

Figure 6. Perspective view of the pilot system and modules; (**1**): rig with pumps, valves and controls, (**2**): module 1; (**3**): module 2; (**4**): feed, concentrate and distillate tanks.

2.4. Key Performance Indicators

For evaluation of the MD process characteristics and comparison with other modules and systems, well-established key performance indicators (KPI) will be introduced in the following and used in this work.

Physical properties of salt solutions differ from pure H_2O. Artificial NaCl–H_2O solutions at different salinities were used for the characterisation of the new module type. Thus, a series of measurements was carried out to allow a conversion of the online measured and logged conductivity values σ (mS/cm) into salinity values S (g/kg). Formulas (1) and (2) were derived for different concentration ranges.

$$S_{(5<\sigma<219.5)} = 0.0017\sigma^2 + 0.4309\sigma + 5.5678 \tag{1}$$

$$S_{(\sigma>219.5)} = 0.0392\sigma^2 + 16.1099\sigma + 1830.2558 \tag{2}$$

The temperature and salinity dependent values for density and specific heat capacity of NaCl–H_2O solutions were derived from the correlations provided by [60].

The weight measurement of the distillate M_d (see Figure 5) over time, allows the calculation of the distillate flow rate $\dot{m}_d$.

From that, the transmembrane Flux j_d (kg/m^2 h) in Equation (3) can be calculated. Here, the distillate flow rate $\dot{m}_d$ is put in relation to the active membrane surface A. Flux is well established in all membrane technologies and therefore allows a good comparison between them regarding the specific production of distillate/permeate.

$$j_d = \frac{\dot{m}_d}{A}\left[\frac{\mathrm{kg}}{(\mathrm{m^2 h})}\right] \tag{3}$$

Another important KPI is the GOR which shows the relation between the thermal energy amount needed for the pure evaporation process of the distillate (numerator) and the amount of heat introduced externally (denominator).

$$GOR = \frac{\dot{m}_d * \Delta h_v}{\dot{m}_{\mathrm{H/C}} * c_\mathrm{p} * (T_\mathrm{hi} - T_\mathrm{co})}[-] \tag{4}$$

wherein h_v (kJ/kg) is the specific evaporation enthalpy, $\dot{m}_{\mathrm{H/C}}$ (kg/h) is the mass flow rate of the heating and cooling solution, $c_{p,H/C}$ (kJ/kgK) the specific heat capacity the heating and cooling solution, T_hi the heating inlet temperature and T_co the cooling outlet temperature. The GOR can be calculated separately for each module by using the matching in- and outlet temperatures or in total for the module pair using the total distillate production of both together with the total heat supplied to the modules.

Within the experiments, distillate mass flow is recorded via a weight value by a balance. Feed flow however, is recorded as a volume flow $\dot{v}_f$. Since the physical properties of NaCl–H_2O solutions change with increasing salinity, the calculation of:

$$\dot{m}_f = \dot{v}_f * \rho_f\left[\frac{\mathrm{kg}}{\mathrm{h}}\right] \tag{5}$$

was carried out with density values determined at the respective feed concentration and 25 °C (temperature at point of volume flow measurement) according to [60]. Based on the same reference, specific heat capacity values c_p were also determined for the respective concentrations for the mean process temperature T_m in the MD module calculated as:

$$T_m = \frac{(T_\mathrm{ci} + T_\mathrm{co} + T_\mathrm{hi} + T_\mathrm{ho})}{4}\ [°\mathrm{C}] \tag{6}$$

wherein T_ci is the temperature at the cooling inlet, T_co the temperature at the cooling outlet and T_hi and T_ho are the temperatures at heating inlet and outlet respectively. T_m was typically between 51–52 °C within the experiments conducted for this work.

Thermal efficiency η_{th} given in Equation (7) describes the percentage share of latent heat for the liquid vapour phase change of distillate in relation to the total heat transported through the membrane.

$$\eta_{th} = \frac{\dot{m}_d \cdot h_v}{\dot{m}_{H/C} \cdot c_{p,H/C} \cdot (T_{hi} - T_{ho})} \cdot 100\ [\%] \tag{7}$$

Recovery ratio (RR) is the ratio of distillate mass flow rate to H/C flow rate:

$$RR = \frac{\dot{m}_d}{\dot{m}_f} \cdot 100\ [\%] \tag{8}$$

In Section 3.3, a new factor R_F is introduced which describes the ratio of heating and cooling flow to feed flow:

$$R_F = \frac{\dot{m}_{H/C}}{\dot{m}_f} [-] \tag{9}$$

3. Results

3.1. Performance Characterization with Tap Water

In order to evaluate and compare the performance of the new module type with the help of KPIs defined in the previous section, measurements and parameter variations with tap water were carried out. The influence of NaCl–H_2O solution on the performance is presented in Section 3.2. The coupling of the module pair, with respect to the serial connection of the feed channels, leads to different performance and temperature profiles in MD 1 and MD 2. The main reason for this is the introduction of additional heat into MD 2, transported by the preheated feed leaving MD 1. Thus, the individual module performance profiles will be analyzed at the beginning of this section, but subsequent evaluation will be carried out for MD 1 and MD 2 combined as a pair, displaying mean values. Volume flow of the heating and cooling loop is given in L/h since it is tap water in all cases and the physical properties, especially density, can be assumed to be constant and equal to 1 kg/L. Regarding the feed flow values, mass flow values were derived from the measured volume flows using physical properties of NaCl solution [60]. The conductivity of the distillate produced was below 19 μS/cm during the entire testing regime with tap water. The pilot system was operated for a total of ~21 weeks with interruptions for the moving of the system from the first pilot site in Berlin to the second site in Freiburg, Germany.

Figure 7 shows the influence of heating and cooling volume flow on flux and GOR in MD 1. For better legibility, flux is depicted on the left and GOR on the right y-axis. Flux increases proportionally with the increase in H/C volume for all three feed mass flows as a result of the higher thermal energy input into the system and subsequent increase in bulk temperature difference. This correlation is a well-known characteristic of the membrane distillation process. The flux values are not very sensitive towards the variation of feed mass flow. However, a certain impact can be observed at 200 and 250 L/h H/C volume flow. When increasing the feed flow, more heat is lost for MD 1 through the heated feed stream. Thus, at lower H/C flow rates at which the overall driving force is lower, the impact of that effect is higher, resulting in lower flux for higher feed mass flow. GOR values for 40 and 50 kg/h of feed follow the well- known trend of reducing with increased H/C volume flow. This is due to that fact, that the increase in driving force temperature difference is not completely compensated by the increase in flux. For a feed mass flow of 50 kg/h the succession of GOR values show that the driving force has been reduced over proportionally by the additional heat requirement to heat the entering feed flow. At 300 L/h H/C flow, the GOR untypically increases slightly due to the now sufficient driving force supplied by the H/C loop. At this point it could be deducted that the feed flow rate should be as low as possible, in order to maximize efficiency. For a single module set up, this conclusion would be valid as long as the saturation level of the feed is not reached within the channel after distillate extraction.

As mentioned earlier, in a double module set-up the heat exiting MD 1 with the feed, is recovered in MD 2 by introducing it into the hot side of the module. The feed stream exits MD 1 at approximately the mean temperature between T_{hi1} and T_{co1} (~69–71 °C) depending on the respective flow profile. Small losses occur through the piping between the two modules. Feed temperature was ~68–70 °C on entering MD 2. The total distillate output of MD 1 was ~10–14 kg/h depending on the flow profile. This corresponds with the flux multiplied by the membrane area of 8 m^2 per module.

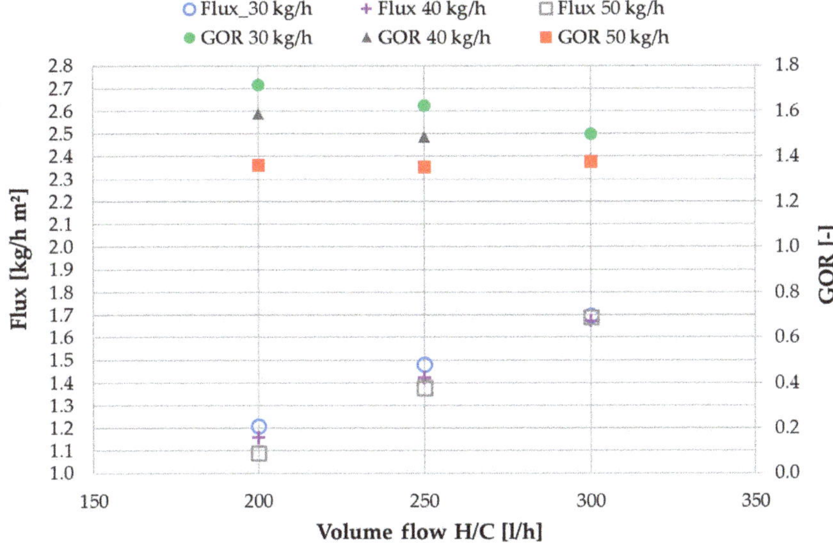

Figure 7. MD 1: influence of H/C volume flow on flux and GOR values at different feed flow rates; tap water; $T_{hi\,1/2}$ = 80 °C, $T_{ci\,1/2}$ = 25.

Figure 7 shows flux and GOR values for MD 2. It must be pointed out, that the feed mass flows entering MD 2 are lower than indicated in the naming 30, 40 and 50 kg/h. These values indicate the mass flow at the inlet of MD 1. Due to the distillate extracted in MD 1, they reach MD 2 reduced by the respective output for each operation point. Table 3 shows the feed gap mass flow values for each H/C flow and FG flow combination tested within the tap water characterization measurements.

Table 3. Feed flow rates at the inlet of MD 1, the inlet of MD 2 and the outlet of MD 2.

H/C (L/h)	FG in M1 (kg/h)	FG in M2 (kg/h)	FG out M2 (kg/h)
400	30	20.2	9.5
400	40	30.7	19.6
400	50	41.2	30.0
500	30	18.1	4.2
500	40	28.5	14.0
500	50	38.9	24.1
600	30	16.3	2.7
600	40	26.5	9.7
600	50	36.4	19.9

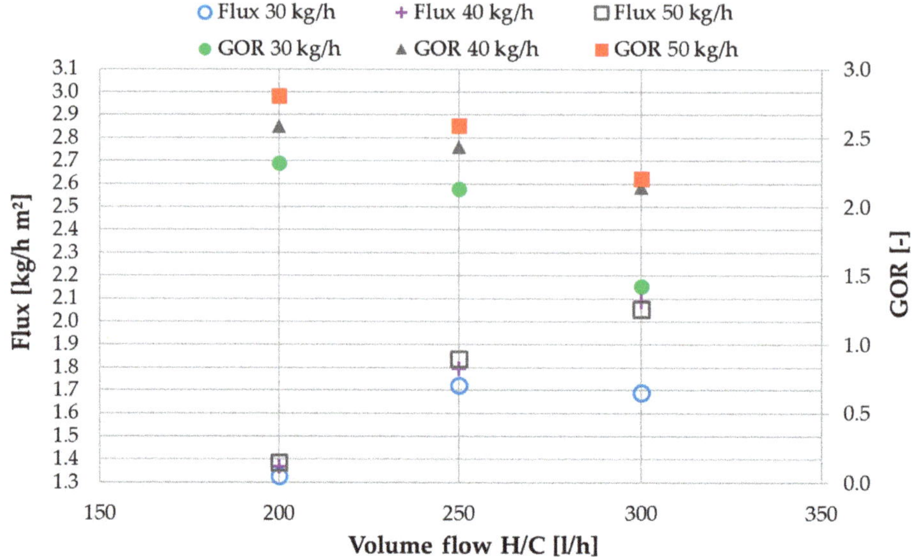

Figure 8. MD 2: influence of H/C volume flow on flux and GOR values at different feed flow rates; tap water; $T_{hi\ 1/2} = 80\ °C$, $T_{ci\ 1/2} = 25$.

There are some major differences in the flux and GOR values of MD 2 which are shown in Figure 8. To begin with, the overall higher flux values, are a direct result of the additional thermal energy being supplied to MD 2 through the preheated feed coming from the outlet MD 1. In MD 2 the flux is largely insensitive to variations of the feed flow with exception of 30 kg/h. At 300 L/h H/C flow and 30 kg/h FG inlet flow, the flux is only 1.7 kg/h m^2 which is equal to the value at a 250 L/h H/C flow. This phenomenon can be explained as follows: In this case the distillate output is limited by the available feed mass flow entering MD 2. Since the flux is the highest at 300 L/h H/C flow and 30 kg/h FG flow in MD 1, not enough feed is transferred to MD 2 for the total possible flux to be produced. As shown in Table 3, 16.3 kg/h are fed to MD 2 which is the lowest FG in MD 2 value overall. Only 2.7 kg/h of feed leave MD 2. Since this phenomenon does not occur at a FG inlet flow of 40 kg/h, it can be concluded that the optimal feed flow for maximized flux at an H/C flow of 600 L/h lies in between 30 and 40 kg/h. The corresponding GOR value is also significantly reduced for that specific operational point since GOR is calculated in relation to the mass of distillate produced.

In opposition to MD 1, in MD 2 the GOR values are the highest for the highest feed flow of 50 kg/h in addition to be being overall higher than the values in MD 1. At a feed flow of 50 kg/h, the GOR is ~3 in MD 2 and ~2 in MD 1 at the same H/C flow of 200 L/h. In analogy to the increased flux, the reason lies in the additional heat supply entering MD 2 through the preheated feed from MD 1. It increases the cooling flow outlet temperature T_{co2}, thus decreasing the delta T (dTh) on the hot side of MD 2. In correspondence with Equation (4) a higher flux or output in the numerator and a lower temperature difference between T_{hi} and T_{co} in the denominator will lead to a higher GOR. This effect increases, the higher the feed flow to MD 2 is, due to the increase in thermal capacity of the flow.

It has been mentioned that the temperature profiles of MD 1 and MD 2 are affected by the feed flow in different degrees. Figure 9 shows the temperature differences dTh on the hot and dTc on the cold side of MD 1 and MD 2 at different H/C volume flows. dTc is the difference between T_{ho} and T_{ci} of the respective module. $T_{hi\ 1/2}$ was set at 80 °C and $T_{ci\ 1/2}$ at 25 °C. The comparison of dT values in Figure 9 is carried out for a feed inlet flow of 40 kg/h. The dTh MD 1 values are higher than then dTc MD 2 values for both the given flow rates. Heat leaves the system with the feed outlet on the hot side of MD 1 leading to a decrease in heat recovery in the module and subsequently a lower T_{co1}. Thus, dTh MD 1 is higher than dT MD 2 in which that heat is added to the thermal energy supply coming from the heat exchanger HEX_{hot}. T_{co2} increases, resulting in the lower dTh MD 2 values depicted in the bar chart. Since the heating inlet temperature T_{hi} is set at 80 °C, the generally increased delta T hot values at 300 L/h are result of decrease in cooling outlet temperature T_{co}. The hot side delta T ratio of MD 1–MD 2 decreases from 4.4 to 3.3 K. Because a higher flux is generated at this flow rate, the feed flow reaching MD 2 is lower than at 200 L/h. This results in a disproportional relation of the hot side delta Ts to one another when the H/C flow rate is increased at the same feed inlet mass flow rate. Opposite effects can be observed regarding the cold side delta Ts. dTc is much lower in MD 1 than in MD 2. On the cold side of the module, T_{ho1} is decreased by the additional thermal capacity of the feed entering MD 1 at approximately the same temperature as T_{ci}. By the time the feed has reached the outlet of MD 2, the mass flow has so far reduced, that the impact on T_{ho2} is low. This is shown by the fact that dTh MD 2 and dTc MD 2 are very similar as could be expected e.g., in a counter current heat exchanger. dTh MD 2 is 0.9 K lower than dTc MD 2 at 200 L/h and 0.8 K lower at 300 L/h H/C flow. It has been shown by [38] that a symmetrical temperature profile along the flow channels of MD modules will have a beneficial effect on process efficiency. This can be achieved by synchronizing the mass flow capacities in the flow channels. In this new channel configuration however, complexity is added by the mass flow capacity of the feed-baring channel. In MD 1, the capacity of the feed channel ads to that of the cooling channel due to their co-current flow relation. In MD 2 however, the feed flows co-current to the heating channel making the temperature profiles in MD 1 and MD 2 not only asymmetrical but also unequal, however, with a much lower impact in MD 2 due to the lower feed mass flow. A reduction in process efficiency must be expected in comparison to a configuration with parallelized temperature profiles as a trade-off for the advantages of being able to establish a one- step process design.

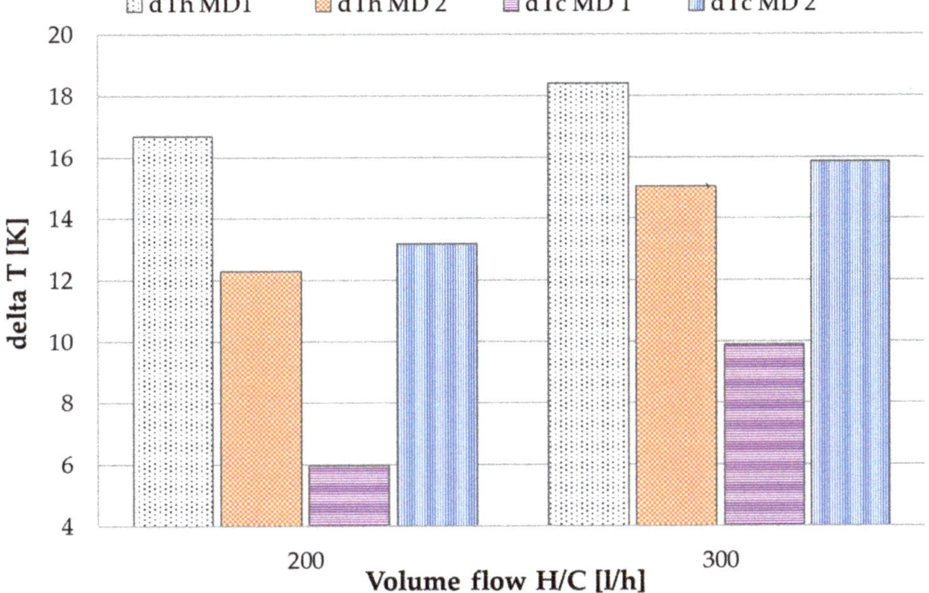

Figure 9. Influence of H/C volume flow on temperature differences in MD 1 and MD 2; $\dot{v}_f = 40$ kg/h, $T_{hi\,1/2} = 80$ °C, $T_{ci\,1/2} = 25$.

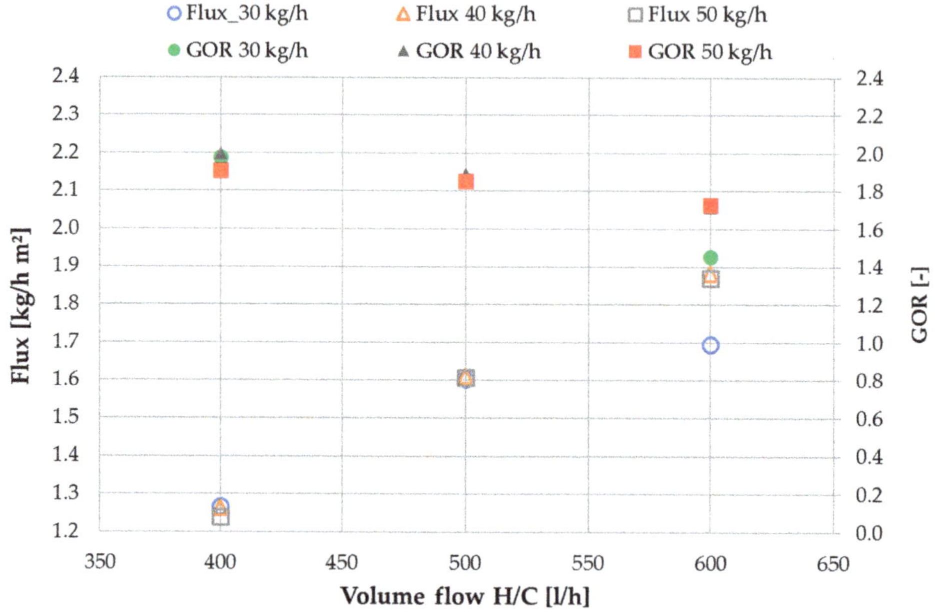

Figure 10. Influence of HC volume flow on Flux and GOR values at varied feed flow rates; MD 1 + MD 2 combined; Tap water; $T_{hi\,1/2} = 80\ °C$, $T_{ci\,1/2} = 25$.

Figure 10 shows mean flux and GOR values for MD 1 and MD 2 combined. Total output was recorded and then divided by the total membrane area of both modules to derive flux. Similarly, GOR was calculated by taking into account the total output and the total heat used for both modules. Total H/C flow for both modules is depicted on the x-axis. It can be observed, that when considering MD 1 and MD 2 as a joined concept, the difference between 40 and 50 kg/h feed flow rate is no longer visible since all sensible heat leaving MD 1 with the feed flow is recovered in MD 2. This is however not valid for 30 kg/h for reasons explained in conjunction with Figure 8. All of the following graphs show combined values for MD 1 and MD 2.

One of the main reasons for separating the heating and cooling flow from the feed flow, is the possibility of controlling the recovery ratio of the module independently of the energy supply. Figure 11 shows the achieved recovery ratios with tap water as feed. Since the recovery ratio RR is a ratio of distillate output to feed input (Equation (8)), more distillate production from the same feed flow rate will result in a higher RR. Thus, with increase of H/C flow the RR values also increase. As shown in Figure 10, the distillate flux does not change significantly with changes in feed flow rate. This results in a reduction of RR for increasing feed mass flow. It should be pointed out, that the highest value of 93% RR was achieved at a H/C flow rate of 600 L/h and feed flow of 30 kg/h. This value might even be exceedable if the aforementioned limitations regarding distillate production in MD 2 had not occurred.

Thermal efficiency η_{th} (Equation (7)) indicates the fraction of latent heat in relation to the total heat transported through the membrane. Since the increase in flux is proportionate to the increase in H/C flow, the η_{th} values presented in Figure 12 remain approximately constant for H/C flow variation. η_{th} will serve as a KPI in the comparison with a previously analyzed spiral wound module in Section 3.4. Since Figure 12 shows mean values for both modules, it is worth mentioning that the individual values were not identical but within ~10% of each other. The tap water characterization presented within this section has shown, that the proposed concept of a FGAGMD channel configuration with a double module strategy provided the desired operational behavior. In the new concept, the basic KPIs flux

GOR, recovery ratio RR and thermal efficiency $\eta_{th.}$ showed the same dependencies as expected in membrane distillation with a very high increase in achievable recovery rate.

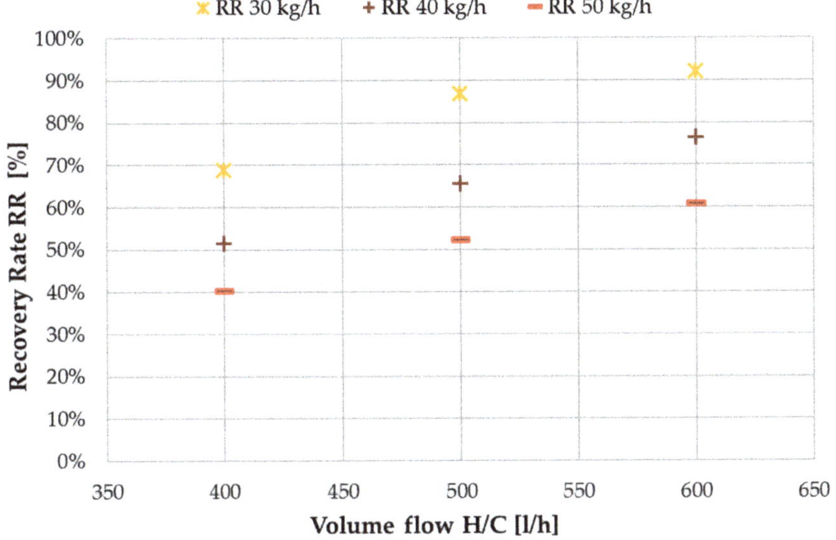

Figure 11. Influence of H/C volume flow on recovery ratio (RR) at different feed flow rates; MD 1 + MD 2 combined; tap water; $T_{hi\ 1/2} = 80$ °C, $T_{ci\ 1/2} = 25$.

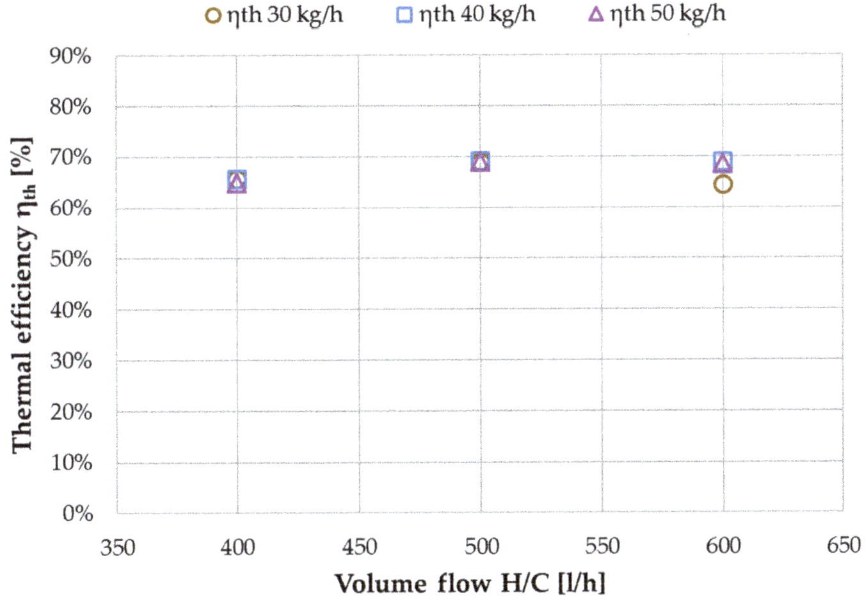

Figure 12. Influence of H/C volume flow on thermal efficiency η_{th} at different feed flow rates; MD 1 + MD 2 combined, tap water, $T_{hi\ 1/2} = 80$ °C, $T_{ci\ 1/2} = 25$.

3.2. Performance Characterization with NaCl Solution

Membrane distillation is capable of concentrating solutions to near saturation, provided that crystal formation does not occur. Being an ambient pressure process, limitations due to e.g., a high osmotic potential of the solution are not an issue. In MD, however, thermal energy is required to bring about the evaporation of the feed solution. With an increasing amount of salt ions in the feed solution, the required energy to evaporate the same amount of distillate increases, due to the decrease in vapour pressure of the solution. For a given MD system with defined channel geometry and fixed operation conditions, this means that flux and GOR will decrease continuously, the more salt ions are present in the feed solution. It is challenging to achieve a high GOR in hypersaline brine concentration with MD. Published data with full-scale modules is scarce but of highest importance for assessment of the real performance of MD technology. Testing and piloting result such as those presented in this work cannot and should not be compared to results achieved with tap water or measured in small scale lab equipment.

Figure 13 shows the value trend of flux in correlation with concentrate salinity at an H/C flow rate of 600 L/h and an inlet feed flow rate of 40 kg/h. The salinity of the concentrate at the outlet of MD 2 was selected for depiction on the x-axis, flux and GOR are shown on the left and right y-axis respectively. During these experiments the blower, indicated in Figure 5, was operated continuously at an avg. air volume flowrate of 17 L/h. It has been shown by [49] in a study with an AGMD module, that when operating with salt solutions, the benefit of optimal draining of the distillate in the air gap channel has a significant positive impact on the distillate quality. A low-pressure air blower is an efficient method of improving the draining. The distillate conductivity was below an avg. of 1 mS/cm during all the salt solution measurements. The well-known impact of vapour pressure reduction [39] can be observed distinctly in Figure 13. Beginning at ~1.9 kg/h for the lowest salinity, flux reduces to 0.87 kg/h for the highest outlet salinity of 214 g NaCl/kg. The succession of GOR values is corresponding to this and provides a range of 1.76–0.78 depending on the concentrate salinity.

The decrease in thermal efficiency η_{th} over salinity is shown in Figure 14. The reduction in vapour pressure caused by the increase in dissolved ions in the solution reduces the effective driving force for evaporation. Thus, the ratio of heat used for actual phase change in relation to the total heat transferred through the membrane and air gap shifts with increasing feed salinity. With tap water, 67% of the total heat is being used for evaporation. At 214 g NaCl/kg this fraction is reduced to 39%. The recovery ratio is furthermore affected by the reduction in flux. At the lowest salinity of 1 g NaCl/kg, 76% of the feed going into the inlet of MD 1 is extracted as distillate. This means that 9.6 kg/h exits MD 2 out of the 40 kg/h which entered MD 1. At 214 g NaCl/kg, the recovery ratio RR reduced to 32% as a result of the reduced flux.

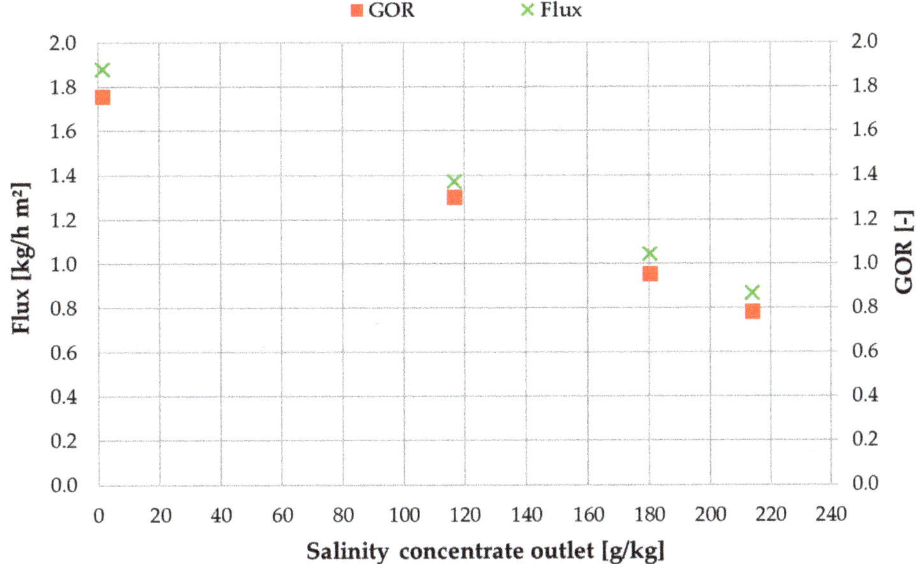

Figure 13. Influence of concentrate salinity on flux and GOR; MD 1 + MD 2 combined; feed inlet salinity 0.3/55/95/143 g NaCl/kg; $\dot{v}_f$ = 40 kg/h, $\dot{v}_{H/C}$ = 600 L/h, $T_{hi\ 1/2}$ = 80 °C, $T_{ci\ 1/2}$ = 25.

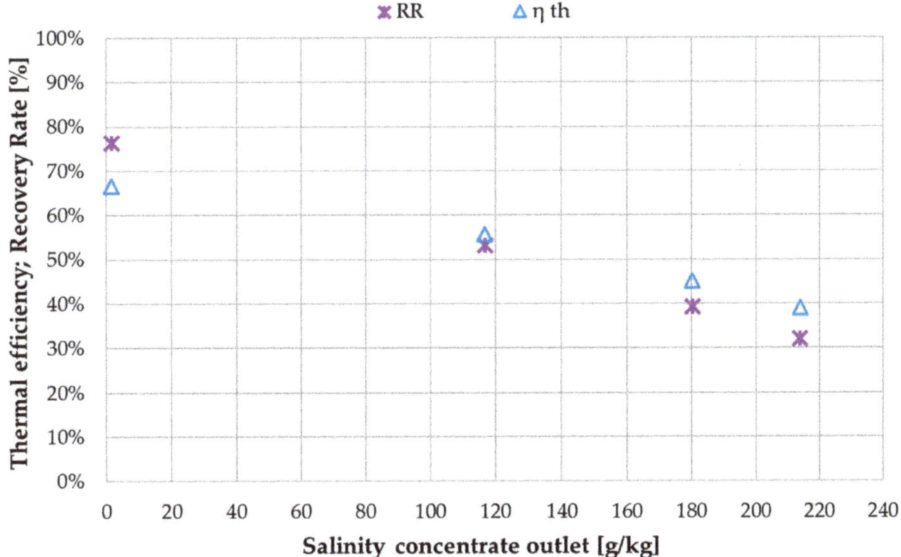

Figure 14. Influence of concentrate salinity on recovery ratio (RR) and thermal efficiency η_{th}; MD 1 + MD 2 combined; Feed inlet salinity 0.3/55/95/143 g NaCl/kg; $\dot{v}_f$ = 40 kg/h, $\dot{v}_{H/C}$ = 600 L/h, $T_{hi\ 1/2}$ = 80 °C, $T_{ci\ 1/2}$ = 25.

3.3. Ratio of H/C Solution Flow to Feed Flow

For a deeper understanding of the benefits of the overall module and system concept some key facts should be elaborated more closely. The goal of the concept is to achieve the desired increase in feed solution concentration in a single pass process. This was not achieved in this particular prototype system due to a restriction in the systems operation. The H/C flowrate was limited to 600 L/h, due to the capacity of pump $P_{H/C}$. As established by the evaluation of the tap water characterization measurements, the ratio of heating and cooling flow rate to feed flow rate has a high impact on the recovery ratio. The effect is connected to the proportionally higher flux production out of the same amount of feed flow when increasing the H/C flow rate. Thus, a new ratio R_F is established (Equation (9)), which describes the quotient of H/C mass flow $\dot{m}_{H/C}$ to feed mass flow $\dot{m}_f$. Figure 15 depicts the influence of R_F on recovery ratio RR for three different feed salinity levels. The impact of both R_F and salinity on recovery ratio can now be observed. Linear fit curves are used to show the progression of the recovery ratio with increasing R_F. The required R_F for a desired recovery ratio increases with increasing feed salinity. For example, for a required RR of 50% in a single pass process at an inlet feed concentration of 1 g NaCl/kg, R_F of ~5 would be sufficient according to the operating temperatures of the pilot system. At 143 g NaCl/kg inlet concentration however, the inclination of the values is much lower and a higher R_F would be required, taking into consideration that NaCl–H_2O solution is saturated at ~253 g NaCl/kg. These correlations are dependent on the module's channel length and are only valid for this specific module. Nonetheless, the principles for future module and process design remain equal for any geometry. For a given inlet feed concentration and a desired final concentration, R_F must be determined in order to achieve the required recovery ratio. Certain boundary conditions must also be considered. In order to sustain overall efficiency, the feed flow cannot be lowered to a value that does not supply a sufficient amount of feed to MD 2. The negative effects on GOR and thermal efficiency were analyzed in Section 3.1. In addition, a safety margin to prevent saturation of the solution in MD 2 should be added to the minimum feed flow rate.

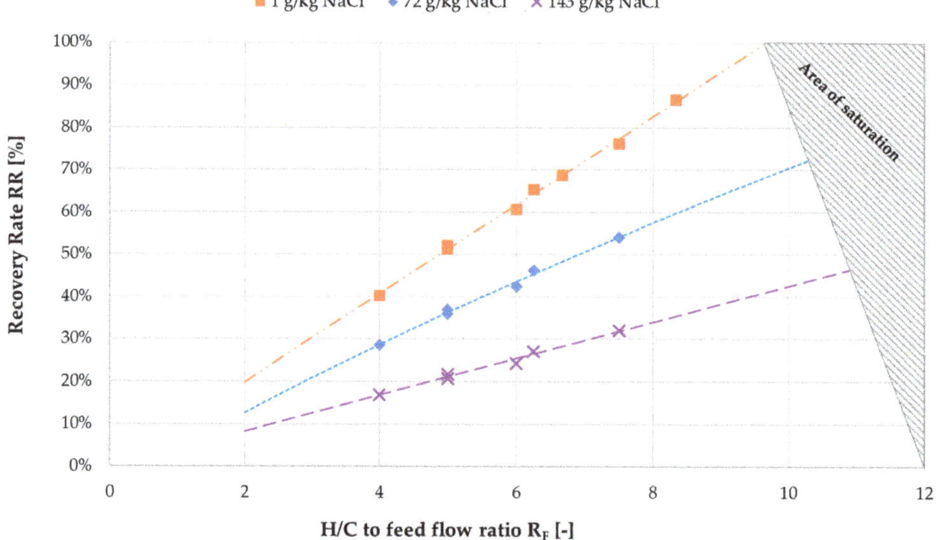

Figure 15. Effect of H/C flow to feed flow ratio R_F on recovery ratio (RR); $T_{hi\,1/2} = 80$ °C, $T_{ci\,1/2} = 25$.

The recommended procedure for selection of feed inlet flow rate and R_F can be summarized as follows:

- Determination of feed concentration;
- Selection of desired outlet concentration;
- Selection of required R_F;
- Determination of minimum feed flow;

- Calculation of H/C flow rate according to R_F.

Depending on the available heat supply, the channel length of the modules will be designed. Channel length selection is a key parameter in MD module design, both technologically as well as economically as explained in detail in [22]. However, the GOR and flux are much more sensitive to channel length modification than the overall output of the module. In consequence, R_F ratios are not expected to change in a large magnitude for the same temperature profile and channel lengths within 4–9 m.

3.4. Comparison with Spiral Wound Air Gap Membrane Distillation (AGMD) Module

Within this work so far, a novel type of MD configuration was presented and analyzed. The analysis was based upon a first-generation prototype plate and frame module. A comparison with the previously studied spiral wound AGMD module type should nonetheless be conducted at this stage in order to identify any possible disadvantages of the overall concept. Even though the packaging of the modules is different, no significant differences in the influence of operational parameters on the KPIs of the process are expected. Materials and membrane types are identical.

The data used for comparison is extracted from a study with an AGMD spiral wound module and hypersaline brine which can be found with [49]. The AGMD module had a 6 m channel length and was operated with the same flow of 300 L/h. Figure 16 shows flux and Figure 17 shows GOR values of both modules in direct comparison. Flux values for both module types are very similar but with slightly higher values for the spiral wound for tap water and slightly higher values for the plate and frame for at salinities above ~80 g NaCl/kg. This shows that for a similar channel length and under the same operating conditions regarding the heating and cooling flows, the flux is similar regardless of the module type. The internal heat recovery of the modules expressed as GOR, however, shows some differences especially at tap water salinity. This effect can be assigned to the additional heat-transfer resistance added by the feed gap in the plate and frame. In the low-salinity region where vapour pressure depreciation does not have an impact, the larger delta T caused by the additional thermal resistance leads to a higher energy requirement per mass unit of distillate which directly effects the GOR. At the same time, this higher effective delta T leads to advantages over the spiral wound module at higher conductivities since more net driving force is available after subtracting the fraction of driving force lost to the reduction in vapour pressure. The asymmetrical temperature profiles in the FGAGMD module shown in Section 3.1. also account for a reduction in the overall GOR values of this module type.

When comparing recovery ratio and thermal efficiency shown in Figure 18, the most significant difference is in the recovery ratio. At tap water the difference is 70% and at 214 g NaCl/kg the difference is still approximately 30%. The enabling of such high recovery ratios was one of the core motivations for the implementation of the new FGAGMD channel configuration. As mentioned previously, the recovery ratio of the spiral wound AGMD module is not independently adjustable due to the coupling of heat supply and feed supply. Thermal efficiency values are similar for both modules with the largest differences at the low end of the salinity range. On average for the entire salinity range tested, however, η_{th} of the FGAGMD module was higher by approximately 4% compared to the spiral wound module.

From the direct comparison of the two module types, it can be deducted that even in this first prototype stage, the plate and frame (PF) FGAGMD module has a general advantage over the spiral wound (SW) AGMD module when implemented in high concentration applications. Under the assumption that e.g., a RO brine at a salinity 7 g NaCl/kg should be concentrated to 240 g NaCl/kg [22], at the resulting avg. salinity of 155 g NaCl/kg the FGAGMD module is be superior in performance regarding flux, η_{th} and RR with only small drawbacks in GOR. Table 4 provides a summary of KPIs at the mentioned avg. salinity of 155 g NaCl/kg.

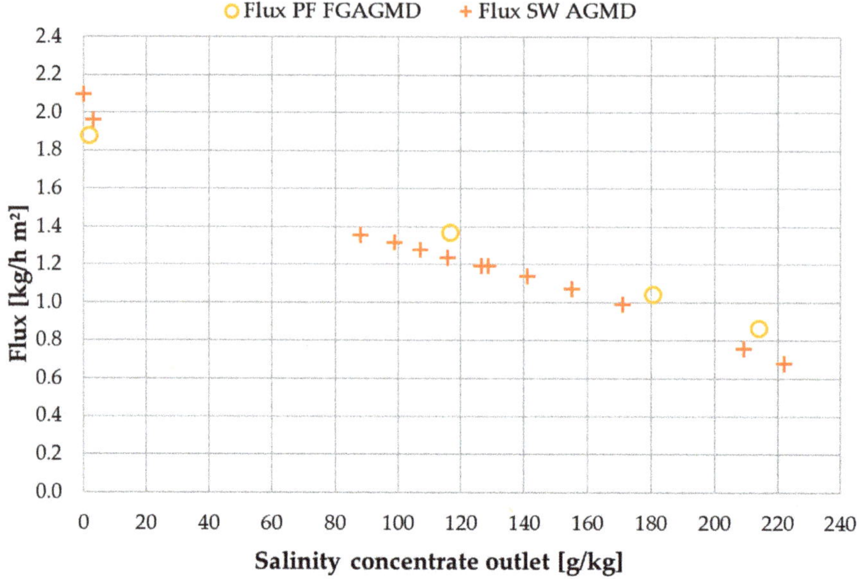

Figure 16. Influence of concentrate salinity on flux in plate and frame FGAGMD and spiral wound AGMD modules; $\dot{v}_f$ = 40 kg/h, $\dot{v}_{H/C}$ = 300 L/h per module; $\dot{v}_{fsw}$ = 300 L/h; T_{hi} = 80 °C, T_{ci} = 25.

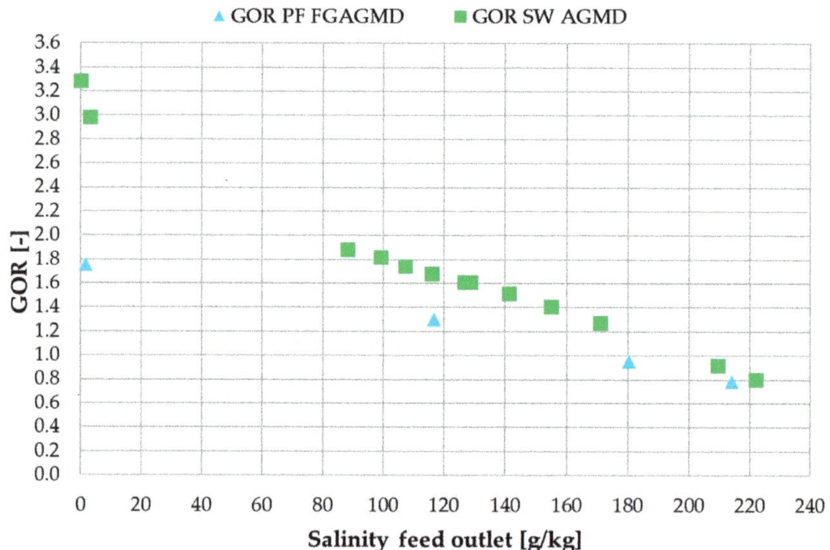

Figure 17. Influence of concentrate salinity on GOR in plate and frame FGAGMD and spiral wound AGMD modules; $\dot{v}_f$ = 40 kg/h, $\dot{v}_{H/C}$ = 300 L/h per module; $\dot{v}_{fsw}$ = 300 L/h; T_{hi} = 80 °C, T_{ci} = 25.

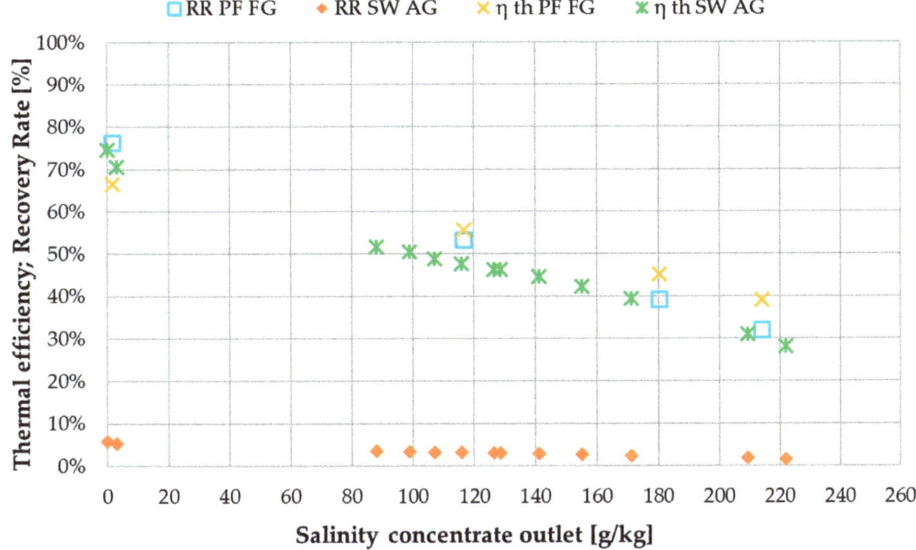

Figure 18. Influence of concentrate salinity on thermal efficiency η_{th} and recovery ratio (RR) in plate and frame and spiral wound module types under similar operating conditions; $\dot{v}_f$ = 40 kg/h, $\dot{v}_{H/C}$ = 300 L/h per module; $\dot{v}_{fsw}$ = 300 L/h; T_{hi} = 80 °C, T_{ci} = 25.

Further potential for improvement on the plate and frame module design is definitely given, especially regarding flow distribution and channel geometry. It is expected that similar GOR values to the AGMD module will be possible after optimization of the inner components of the module. Furthermore, an increase in channel length can be considered in order to increase the GOR. This will, however, decrease the flux and require more membrane area. The application of vacuum to the air gap would have a significant positive effect on efficiency, though drawbacks in distillate quality are to be expected when implementing this option.

Table 4. Average key performance indicator (KPI) values for plate and frame FGAGMD and spiral wound AGMD module, mean concentrate salinity ~155 mS/cm; $\dot{v}_f$ = 40 kg/h, $\dot{v}_{H/C}$ = 300 L/h per module; $\dot{v}_{fsw}$ = 300 L/h; T_{hi} = 80 °C, T_{ci} = 25.

	Flux (kg/m^2 h)	GOR (-)	RR (%)	ηth (%)
PF FGAGMD	1.2	1.1	45	50
SW AGMD	1.1	1.4	3	46

4. Conclusions

Within this work, a novel plate and frame FGAGMD module was presented. By separating the heating and cooling channel from the feed channel, the concept allows a minimum amount of components in contact with the highly corrosive feed. Furthermore, it decoupled the thermal energy supply from the feed supply, giving room for a new range of operational flow settings. The goal of increasing the recovery ratio of the single pass process was achieved with values of up to 93% RR using tap water as feed and between 32–53% with NaCl solutions ranging between 117 and 214 g NaCl/kg. The impact of well-known correlations on the KPIs flux, GOR and η_{th} remained valid for this new module type. For optimization of flow rate strategies for a given range of concentration a new ratio of H/C flow to feed flow was introduced named R_F. R_F serves as an indicator for the selection of a flow regime to achieve a required recovery ratio. In comparison to a previously analyzed AGMD spiral wound module, the FGAGMD plate and frame prototype showed similar performance characteristics with slight improvements regarding flux and η_{th}. As expected, the recovery ratio was between 12–16 times higher in the plate and frame module (32–93%) than in the spiral wound (2–6%) due to the new channel configuration. Drawbacks of the FGAGMD module were observed in GOR especially in the lower salinity region. For an average salinity of 155 mS/cm, however, the difference in GOR reduced approx. 0.3. It is likely that the GOR of the plate and frame module can be improved by optimizing the internal flow distribution within the channels which was not the focus in the construction of this first prototype module. In general, for the applications with corrosive, toxic or otherwise hazardous media the implementation of a FGAGMD plate and frame MD module opens up a new range of applications for the technology with advantages in cleaning and maintenance, safety and the integration into industrial wastewater treatment processes. Further improvements must be carried out to optimize the efficiency and long- term testing will be necessary with a follow-up prototype in order to properly evaluate lifetime cycles of all replacement parts.

Furthermore, an in depth investigation of the thermodynamics of the FGAGMD process in a flat sheet bench scale testing facility will published in a follow-up publication. This will enable the validation of modelling tools and the optimization of efficiency and operational key factor R_F with respect to different salinity levels.

Author Contributions: R.S. was the lead investigator and author of the original manuscript. She provided the methodology, and final data evaluation and managed and supervised the pilot operation. The prototype module was conceptualized by R.S. and L.B. L.B. constructed and built the prototype. J.S. operated the prototype, conducted preliminary data evaluation and was supervised by R.S. and T.H. Expert input regarding the methodology, and draft review was provided by D.W. and J.K. Further reviewing and editing was conducted by T.H. and S.-U.G.

Funding: This research was funded by the German Federal Ministry of Education and Research, grant number 02WAV1406E.

Conflicts of Interest: The authors declare no conflict of interest. The founding sponsors had no role in the design of the study; in the collection, analyses, or interpretation of data; in the writing of the manuscript; and in the decision to publish the results.

Nomenclature

A	Membrane area
AGMD	Air gap membrane distillation
C	Conductivity (mS/cm; µS/cm)
cp	Specific heat capacity (kJ/kg K)
DCMD	Direct contact membrane distillation
delta T	Driving force temperature difference (K)
dTh	Temperature difference on the hot side of the module
dTc	Temperature difference on the cold side of the module
F	Flow (L/h)
FGMD	Feed Gap Membrane Distillation
FGAGMD	Feed Gap Air Gap Membrane Distillation
GOR	Gained output ratio (-)
h	Height (m)
HEX	Heat exchanger
KPI	Key performance indicator
m	Mass flow (kg/h)
MD	Membrane Distillation
MED	Multi Effect Distillation
MSF	Multi Stage Flash
j	Transmembrane flux (kg/m^2 h)
L	Channel length (m)
p	Pressure (bar)
P	Pump

PP	Polypropylene
PTFE	Polytetrafluorethylene
PGMD	Permeate gap membrane distillation
R	Ratio
RR	Recovery ratio
RO	Reverse Osmosis
S	Salinity (g/kg)
SWRO	Seawater Reverse Osmosis
T	Temperature (°C)
ρ	Density (kg/m^3]
η	Efficiency (%)
v-AGMD	Vacuum air gap membrane distillation
v	Volume flow (L/h)
Δhv	Evaporation enthalpy (kJ/kg)

Indices

ci	cooling inlet
co	cooling outlet
d	distillate
hi	heating inlet
ho	heating outlet
f	feed
F	feed ratio
H/C	heating and cooling
hx	heat exchanger
m	mean
th	thermal

References

1. Criscuoli, A.; Drioli, E.; Capuano, A.; Memoli, B.; Andreucci, V. Human plasma ultrafiltrate purification by membrane distillation: process optimisation and evaluation of its possible application on-line. *Desalination* **2002**, *147*, 147–148. [CrossRef]
2. Ding, Z.; Liu, L.; Li, Z.; Ma, R.; Yang, Z. Experimental study of ammonia removal from water by membrane distillation (MD): The comparison of three configurations. *J. Membr. Sci.* **2006**, *286*, 93–103. [CrossRef]
3. Grzechulska-Damszel, J.; Tomaszewska, M.; Morawski, A. Integration of photocatalysis with membrane processes for purification of water contaminated with organic dyes. *Desalination* **2009**, *241*, 118–126. [CrossRef]
4. Khayet, M. Treatment of radioactive wastewater solutions by direct contact membrane distillation using surface modified membranes. *Desalination* **2013**, *321*, 60–66. [CrossRef]
5. Ma, R.; El-Bourawi, M.; Khayet, M.; Ding, Z.; Li, Z.; Zhang, X. Application of vacuum membrane distillation for ammonia removal. *J. Membr. Sci.* **2007**, *301*, 200–209.
6. Sarti, B. Concentration of must through vacuum membrane distillation. *Desalination* **2002**, *149*, 253–259.
7. Lagana, F.; Barbieri, G.; Drioli, E. Direct contact membrane distillation: modelling and concentration experiments. *J. Membr. Sci.* **2000**, *166*, 1–11. [CrossRef]
8. Hitsov, I.; Maere, T.; de Sitter, K.; Dotremont, C.; Nopens, I. Modelling approaches in membrane distillation: A critical review. *Sep. Purif. Technol.* **2015**, *142*, 48–64. [CrossRef]
9. Winter, D. Membrane Distillation—A Thermodynamic, Technological and Economic Analysis. Ph.D. Thesis, University of Kaiserslautern, Rhineland-Palatinate, Germany, 2015. Available online: http://www.reiner-lemoine-stiftung.de/pdf/dissertationen/Dissertation-Winter.pdf (accessed on 6 September 2019).
10. Eykens, L.; de Sitter, K.; Dotremont, C.; Pinoy, L.; van der Bruggen, B. Characterization and performance evaluation of commercially available hydrophobic membranes for direct contact membrane distillation. *Desalination* **2016**, *392*, 63–73. [CrossRef]
11. Li, X.; Shan, H.; Cao, M.; Li, B. Facile fabrication of omniphobic PVDF composite membrane via a waterborne coating for anti-wetting and anti-fouling membrane distillation. *J. Membr. Sci.* **2019**, *589*, 117262. [CrossRef]

12. Fane, A.; Tun, C.; Matheickal, J.; Sheikholeslami, R. Membrane distillation crystallisation of concentrated salts—Flux and crystal formation. *J. Membr. Sci.* **2005**, *257*, 144–155.
13. Arafat, A.; Saffarini, R.; Mansoor, B.; Thomas, R. Effect of temperature-dependent microstructure evolution on pore wetting in PTFE membranes under membrane distillation conditions. *J. Membr. Sci.* **2013**, *429*, 282–294.
14. Guillen-Burieza, E.; Mavukkandy, M.; Bilad, M.R.; Arafat, H. Understanding wetting phenomena in membrane distillation and how operational parameters can affect it. *J. Membr. Sci.* **2016**, *515*, 163–174. [CrossRef]
15. Guillen-Burrieza, E. Membrane structure and surface morphology impact on the wetting of MD membranes. *J. Membr. Sci.* **2015**, *483*, 94–103. [CrossRef]
16. Lienhard, J.; Warsinger, D.; Servi, A.; Connors, G.; Mavukkandy, M.; Arafat, H.; Gleason, K. Reversing membrane wetting in membrane distillation: comparing dryout to backwashing with pressurized air. *Environ. Sci. Water Res. Technol.* **2017**, *3*, 930–939.
17. Peng, Y.; Ge, J.; Li, Z.; Chen, P.; Wang, S. Membrane fouling and wetting in a DCMD process for RO brine concentration. *Desalination* **2014**, *344*, 97–107.
18. Rezaei, M.; Warsinger, D.M.; Lienhard, J.H.; Samhaber, W.M. Wetting prevention in membrane distillation through superhydrophobicity and recharging an air layer on the membrane surface. *J. Membr. Sci.* **2017**, *530*, 42–52. [CrossRef]
19. Rezaei, M.; Warsinger, D.; Lienhard, J.; Duke, M.; Matsuura, T.; Samhaber, W. Wetting phenomena in membrane distillation: Mechanisms, reversal, and prevention. *Water Res.* **2018**, 329–352. [CrossRef] [PubMed]
20. Nashed, A. High recovery rate solar driven reverse osmosis and membrane distillation plants for brackish groundwater desalination in Egypt. Ph.D. Thesis, University of New South Wales, Sydney, Australia, 2014.
21. Volpin, F.; Chekli, L.; Ghaffour, N.; Vrouwenvelder, J.; Shon, H.K. Optimisation of a forward osmosis and membrane distillation hybrid system for the treatment of source-separated urine. *Sep. Purif. Technol.* **2019**, *212*, 368–375. [CrossRef]
22. Schwantes, R.; Chavan, K.; Winter, D.; Felsmann, C.; Pfafferott, J. Techno-economic comparison of membrane distillation and MVC in a zero liquid discharge application. *Desalination* **2018**, *428*, 50–68. [CrossRef]
23. Tavakkoli, S.; Lokare, O.; Vidic, R.; Khanna, V. A techno-economic assessment of membrane distillation for treatment of Marcellus shale produced water. *Desalination* **2017**, *416*, 24–34. [CrossRef]
24. Al-Asheh, S.; Banat, F.; Qtaishat, M.; Al-Khateeb, M. Concentration of sucrose solutions via vacuum membrane distillation. *Desalination* **2006**, *195*, 60–68. [CrossRef]
25. Hilal, N.; Gunko, S.; Verbych, S.; Bryk, M. Concentration of apple juice using direct contact membrane distillation. *Desalination* **2006**, *190*, 117–124.
26. Macedonio, F.; Quist-Jensen, C.; Coindi, C.; Cassano, A.; Aljlil, S.; Alharbi, O.; Drioli, E. Direct contact membrane distillation for the concentration of clarified orange juice. *J. Food Eng.* **2016**, *187*, 37–43.
27. Alkhudhiri, A. Membrane Distillation: A comprehensive review. *Desalination* **2012**, *287*, 1–7. [CrossRef]
28. Drioli, E. Membrane distillation: Recent developments and perspectives. *Desalination* **2015**, *356*, 56–84. [CrossRef]
29. Warsinger, D.M. Scaling and fouling in membrane distillation for desalination applications: A review. *Desalination* **2015**, *365*, 294–313. [CrossRef]
30. Choudhury, M.; Anwar, N.; Jassby, D.; Rahaman, M. Fouling and wetting in the membrane distillation driven wastewater reclamation process—A review. *Adv. Colloid Interface Sci.* **2019**, *269*, 370–399. [CrossRef]
31. Singh, D.; Li, L.; Obuscovic, G.; Chau, J.; Sirkar, K. Novel Cylindrical Cross-Flow Hollow Fiber Membrane Module for Direct Contact Membrane Distillation-based Desalination. *J. Membr. Sci.* **2018**, *545*, 312–322. [CrossRef]
32. Song, L.; Ma, Z.; Liao, X.; Kosaraju, P.; Irish, J.; Sirkar, K. Pilot plant studies of novel membranes and devices for direct contact membrane distillation-based desalination. *J. Membr. Sci.* **2008**, *232*, 257–270. [CrossRef]
33. He, F.; Gilron, J.; Sirkar, K. High water recovery in direct contact membrane distillation using a series of cascades. *Desalination* **2013**, *323*, 48–54. [CrossRef]
34. Singh, D.; Sirkar, K. Desalination by air gap membrane distillation using a two hollow-fiber-set membrane module. *J. Membr. Sci.* **2012**, *421–422*, 172–179. [CrossRef]

35. Hitsov, I.; de Sitter, K.; Dotremont, C.; Cauwenberg, P.; Nopens, I. Full-scale validated Air Gap Membrane Distillation (AGMD) model without calibration parameters. *J. Membr. Sci.* **2017**, *533*, 309–320. [CrossRef]
36. Lee, H.; He, F.; Song, L.; Gilron, J.; Sirkar, K. Desalination with a Cascade of Cross-Flow Hollow Fiber Membrane Distillation Devices Integrated with a Heat Exchanger. *AIChE* **2011**, *57*, 1780–1795. [CrossRef]
37. Gilron, J.; Song, L.; Sirkar, K. Design for Cascade of Crossflow Direct Contact Membrane Distillation. *Ind. Eng. Chem. Res.* **2007**, *46*, 2324–2334. [CrossRef]
38. Winter, D.; Koschikowski, J.; Gross, F.; Maucher, D.; Düver, D.; Jositz, M.; Mann, T.; Hagedorn, A. Comparative analysis of full-scale membrane distillation contactors—Methods and modules. *J. Membr. Sci.* **2017**, *524*, 758–771. [CrossRef]
39. Alkhudhiri, A.; Hilal, N. Air gap membrane distillation: A detailed study of high saline solution. *Desalination* **2017**, *403*, 179–186. [CrossRef]
40. Schofield, R.; Fane, A.; Fell, C. Heat and mass transfer in membrane distillation. *J. Membr. Sci.* **1987**, *33*, 299–313. [CrossRef]
41. Bodell, B. Distillation of saline water using silicone rubber membrane. U.S. Patent Patent 3361645A, 3 June 1963.
42. Wang, P.; Chung, T.-S. Recent advances in membrane distillation processes: Membrane development, configuration design and application exploring. *J. Membr. Sci.* **2015**, *474*, 39–56. [CrossRef]
43. Yang, X.; Wang, R.; Fane, A. Novel designs for improving the performance of hollow fiber membranedistillation modules. *J. Membr. Sci.* **2011**, *384*, 52–62. [CrossRef]
44. Eykens, L.; Reyns, T.; de Sitter, K.; Dotremont, C.; Pinoy, L.; van der Bruggen, B. How to select a membrane distillation configuration? Process conditions and membrane influence unraveled. *Desalination* **2016**, *399*, 105–115. [CrossRef]
45. Ruiz-Aguirre, A.; Andres-Manas, J.; Fernandes-Sevilla, J.; Zaragoza, G. Experimental characterization and optimization of multi-channel spiral wound air gap membrane distillation modules for seawater desalination. *Sep. Purif. Technol.* **2018**, *205*, 212–222. [CrossRef]
46. Memsys GmbH. Available online: Available:http://www.memsys.eu/ (accessed on 10 March 2016).
47. Heinzl, W.; Zhao, K.; Wenzel, M.; Büttner, S.; Bollen, F.; Lange, G.; Heinzl, S.; Sarda, N. Experimental study of the memsys vacuum multi effect membrane distillation (V-MEMD) module. *Desalination* **2013**, *323*, 150–160.
48. Eykens, L.; Hitsov, I.; de Sitter, K.; Dotremont, C.; Pinoy, L.; van der Bruggen, B. Direct contact and air gap membrane distillation: Differences and similarities between lab and pilot scale. *Desalination* **2017**, *422*, 91–100. [CrossRef]
49. Schwantes, R.; Bauer, L.; Chavan, K.; Dücker, D.; Felsmann, C.; Pfafferott, J. Air gap membrane distillation for hypersaline brine concentration: Operational analysis of a full-scale module–New strategies for wetting mitigation. *Desalination* **2018**, *444*, 13–25. [CrossRef]
50. Schwantes, R.; Cipollina, A.; Gross, F.; Koschikowski, J.; Pfeifle, D.; Subiela, V.; Rolletschek, M. Membrane distillation: Solar and waste heat driven demonstration plants for desalination. *Desalination* **2013**, *323*, 93–106. [CrossRef]
51. Raluy, G.; Schwantes, R.; Ortin, V.S.; Penate, B.; Melian, G.; Betancort, J. Operational experience of a solar membrane distillation demonstration plant in Pozo Izquierdo-Gran Canaria Island (Spain). *Desalination* **2012**, *290*, 1–13. [CrossRef]
52. Winter, D.; Koschikowski, J.; Wieghaus, M. Desalination using membrane distillation: Experimental studies on full scale membrane distillation modules. *J. Membr. Sci.* **2011**, *375*, 104–112. [CrossRef]
53. Koschikowski, J.; Wieghaus, M.; Rommel, M.; Ortin, V.S.; Penate, B.; Betancort, J. Experimental investigations on solar driven stand-alone membrane distillation systems for remote areas. *Desalination* **2009**, *248*, 125–131. [CrossRef]
54. Banat, F.; Jwaied, N.; Rommel, M.; Koschikowski, J.; Wieghaus, M. Desalination by a "compact SMADES" autonomous solarpowered membrane distillation unit. *Desalination* **2007**, *217*, 29–37. [CrossRef]
55. Swaminathan, J.; Chung, H.; Warsinger, D.; Lienhard, J. Energy efficiency of membrane distillation up to high salinity: Evaluating critical system size and optimal membrane thickness. *Appl. Energy* **2018**, *211*, 715–734. [CrossRef]
56. Hausmann, A.; Sanciolo, P.; Vasiljevic, T.; Weeks, M.; Duke, M. Membrane distillation in the dairy industry: process integration and membrane performance. In Proceedings of the International Workshop on Membrane Distillation and Related Technologies, Ravello, Italy, 9–12 October 2011; pp. 93–96.

57. Hausmann, A.; Sanciolo, P.; Vasiljevic, T.; Weeks, M.; Duke, M. Integration of Membrane Distillation into Heat Paths of Industrial Processes. *Chem. Eng. J.* **2012**, *211–212*, 378–387. [CrossRef]
58. El-Bourawi, M.; Ding, Z.; Ma, R.; Khayet, M. A framework for better understanding membrane distillation separation process. *J. Membr. Sci.* **2006**, *285*, 4–29. [CrossRef]
59. Khayet, M. Membranes and theoretical modeling of membrane distillation: A review. *Adv. Colloid Interface Sci.* **2011**, *164*, 56–88. [CrossRef] [PubMed]
60. Ramalingam, A.; Arumugam, S. Experimental Study on Specific Heat of Hot Brine for Salt Gradient Solar Pond Application. *Int. J. Chemtech Res.* **2012**, *4*, 56–961.

Article

The Application of Submerged Modules for Membrane Distillation

Marek Gryta

Faculty of Chemical Technology and Engineering, West Pomeranian University of Technology in Szczecin, ul. Pułaskiego 10, 70-322 Szczecin, Poland; marek.gryta@zut.edu.pl

Received: 20 December 2019; Accepted: 4 February 2020; Published: 6 February 2020

Abstract: This paper deals with the efficiency of capillary modules without an external housing, which were used as submerged modules in the membrane distillation process. The commercial hydrophobic capillary membranes fabricated for the microfiltration process were applied. Several constructional variants of submerged modules were discussed. The influence of membrane arrangement, packing density, capillary diameter and length on the module performance was determined. The effect of process conditions, i.e., velocity and temperature of the streams, on the permeate flux was also evaluated. The submerged modules were located in the feed tank or in the distillate tank. It was found that much higher values of the permeate flux were obtained when the membranes were immersed in the feed with the distillate flowing inside the capillary membranes. The efficiency of submerged modules was additionally compared with the conventional membrane distillation (MD) capillary modules and a similar performance of both constructions was achieved.

Keywords: membrane distillation; submerged module; capillary membrane

---

1. Introduction

The membrane processes are carried out in the membrane modules, usually composed of the membranes assembled in a tubular housing. Such a configuration enables the desired cross-flow velocity under different conditions of pressure and flow rate in order to control the concentration polarization and to achieve the appropriate values of the transmembrane pressure (TMP). In Japan (1988–1989) an idea was presented in which the bundles of submerged hollow fibres were assembled inside a non-pressure vessel and the filtration was realized by the removal of permeate by suction [1]. This low-pressure variant of micro- and ultrafiltration (MF and UF) was commercialised and applied in the water and wastewater industry [1–3].

The submerged modules concept is well established for the filtration of solutions containing a substantial quantity of suspended solids. The low values of TMP (usually below 0.05 MPa) limit fouling through the achievement of operating conditions close to the critical flux [4,5]. With regard to this, the submerged modules are employed in the membrane bioreactors (MBR) [6–8]. In this case, the membranes are usually assembled in the cassette system with feeding of air from the bottom of the cassette [1,8]. Air bubbling induces the fibre movement and the flow of feed along the membranes, which generates the surface shear, mitigating the deposition of a fouling layer [1,7]. Moreover, a periodic backwashing and additional chemical cleaning is employed for the removal of deposits from the membrane surface, thus enabling its long-term exploitation [4,7–10].

In membrane distillation (MD) the water evaporates through the pores of nonwetted, hydrophobic membranes. In this process, the driving force for mass transfer is a vapour pressure difference, but not the TMP. Therefore, the reasons for application of the submerged modules in the MD process are different than those in the case of MF or UF processes. The permeate collected during the single flow of the feed through the MD module constitutes only a few percent of feed volume filling the module channels. With regard to this, in order to achieve a higher degree of water recovery, the feed is circulated in the closed loop and simultaneously reheated before the feed water enters the module [11,12]. These operations are omitted by immersing the capillary membranes inside the feed tank. Thus, the application of submerged modules significantly simplifies a construction of installation, which also additionally reduces the heat losses in the MD process [13–15].

During a long-term exploitation of the MD module, the water fills a fraction of the membrane pores, thereby the elimination of membrane wettability becomes a very important issue [16–18]. The MD process is not pressure-driven, however, the hydrostatic pressure is necessary in order to obtain the feed flow through the module channels. Although the value of this pressure is usually not high, this is one of the reasons accelerating the membrane wetting, especially if the surface tension of the feed solution is low [17,18]. The immersion of membranes inside a non-pressure feed vessel allows elimination of a hydrostatic pressure generated by feed pumping.

Fouling is also one of the reasons causing the membrane wetting [19–21]. The application of air bubbling is commonly employed in the MBR in order to mitigate the fouling phenomenon and it also proves to be effective in the MD process [15,22,23]. Moreover, in the case of the presence of surface active agents, they are accumulated on the surface of air bubbles, which additionally limits the influence of these substances on the pore wetting [23]. Such a concept was successfully used in the MD process applied to the treatment of wastewater from the petrochemical industry and to the treatment of produced water generated during the extraction of gas and petroleum [2,24].

The driving force in the MD process does not directly depend on the bulk feed temperature, but on the feed/membrane interfacial temperature [25–31]. Therefore, the hydrodynamic conditions are important since the occurrence of temperature polarization reduces in a significant degree the MD process efficiency [30]. Bubble aeration and mixing allow the cross-flow of feed through the membrane bundle [32], however, it is difficult to obtain such a high flow turbulence in the submerged modules as in the case of the conventional cross-flow module with housing [30,33,34]. Using a Vacuum Membrane Distillation (VMD) variant can enhance the values of the driving force in this case [26], however, the application of pressure difference (vacuum) also accelerates the membrane wetting [13,30]. On that account the submerged modules working in a Direct Contact MD (DCMD) variant were utilized in the presented work.

Due to the heat and mass transfer, the feed temperature in the boundary layer decreases rapidly, which results in a significant decline of the permeate flux. On that account, the turbulent/mixing conditions are important, especially when a higher feed temperature is used [35]. The submerged modules in the capillary configuration possess only one flow channel in the lumen side. The distillate usually flows through this channel; therefore, a movement (e.g., stirring) of the liquid filling the feed tank can decrease the negative effect of temperature polarization on the feed side. In this case, the feed flow in the submerged module system can be induced by the air bubbles introduced between the fibres. The effect of liquid movement in the feed tank can be also obtained by mechanical mixing [32] or by pumping of feed (recirculation) [36].

An enhancement of flow turbulence on the feed side limits the negative effect of the temperature polarization [34–40]. However, the bulk temperature is also decreased during the MD process, hence, the achievement of higher degrees of water recovery will require the energy supply also in the case of application of the submerged MD modules. The location of elements for heat transfer directly under the modules in the feed tank is advantageous. Such a concept limits the heat losses and enhances the feed flow turbulence between the membranes through the formation of the convection currents. In this case, a high effectiveness of heat transfer for the cross-flow system can be achieved by assembling the membranes in a horizontal position. Other concepts include the feed circulation in a loop from the feed tank to the external heat exchangers. In such a case, a significant influence on the process effectiveness includes a kind of energy source and a connection method of the MD installation with heat exchangers [41].

The advantages of MD submerged modules render them suitable for the application in MBR or in the MD process integrated with crystallization (MDC) and forward osmosis [27]. The studies of MD submerged modules presented in the literature were conducted using the capillary membranes of short length, usually not exceeding 20–30 cm. Therefore, a further development of the MD process requires additional testing of these modules also on a larger scale [26,34]. The submerged modules fabricated from capillary polypropylene membranes are currently offered by the ZENA Membranes company from the Czech Republic [42]. Besides the smaller laboratory modules, the company also produces larger submerged modules with a working membrane area of over 4 m^2, which can be utilized for building the MD pilot installations. The size of the modules, especially the length of the capillaries and their packing density, has a decisive impact on the MD process. For this reason, the effect of module construction, capillary membrane arrangement and location in the feed tank, and the MD process conditions on the module performance was examined in the presented work.

2. Heat and Mass Transfer

In the DCMD variant, a hydrophobic membrane separates the hot feed from the cooled distillate (Figure 1). The water vapour diffuses through the gas phase, filling the membrane pores, and a difference of vapour partial pressure (Δp) creates the driving force for mass transfer. The value of Δp depends on the temperature of the evaporation surface (T_1) and condensation surface (T_2).

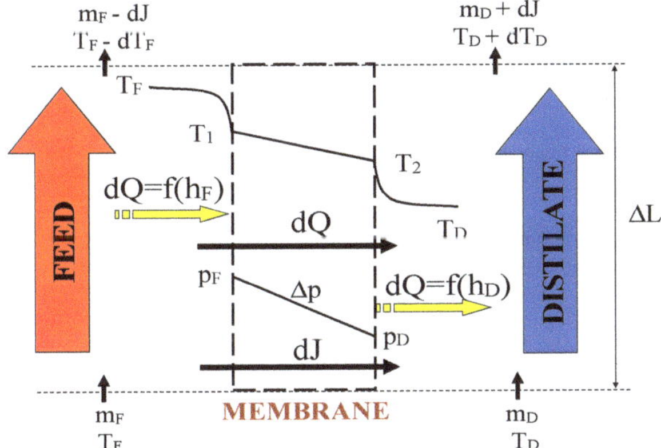

Figure 1. Scheme of heat and mass transfer in membrane distillation (MD) process. m_i—mass flow rate [kg/s].

The vapour pressure p_i of water for the diluted solutions can be determined from the Antoine equation [43]:

$$\log p_i = 8.07131 - \frac{1730.63}{233.42 + t_i} \tag{1}$$

where p_i [mmHg] and t_i—temperature (°C).

The effect of membrane morphology and the operating conditions of the MD process on the magnitude of the obtained permeate flux (J) is often described by equation [44]:

$$J = \frac{\varepsilon}{\chi s} \frac{M}{RT_m} D_{WA} P \ln \frac{P - p_D}{P - p_F} \tag{2}$$

where p_F and p_D are the partial pressures of the saturated water vapour at the interfacial temperatures T_1 and T_2, on the feed and distillate side, respectively; ε—porosity, χ—tortuosity (in this work was assumed χ = 2), and s—membrane thickness. Moreover, P is process pressure, M—molecular mass,

R—gas constant, T_m—average membrane temperature, and D_{WA}—the effective diffusion coefficient of water vapour through the membrane pores.

The value D_{WA} was estimated based on the molecular (D_M) and Knudsen (D_K) diffusion coefficients:

$$\frac{1}{D_{WA}} = \frac{1}{D_M} + \frac{1}{D_K} \tag{3}$$

where:

$$D_K = \frac{d_P}{3}\sqrt{\frac{8RT_m}{\pi M}} \tag{4}$$

$$D_M = D\left(\frac{T_m}{273.15}\right)^{1.75} \tag{5}$$

and water vapour diffusion in air (D) is equal to 0.297×10^{-4} m^2/s [45].

The interfacial temperatures T_1 and T_2 are different from the measured bulk temperatures on the distillate and feed side. This phenomenon is called the temperature polarization, the occurrence of which reduces the value of the driving force for mass transfer in the MD process [39,40]. The temperatures T_1 and T_2 were calculated using numerical modelling of the MD process [26,27]. In the presented work, a model described in the previous papers [44,46] was used for MD process simulations. In this model, the interfacial temperatures were calculated using the following equations:

$$T_1 = \frac{k_m\left(T_D + \frac{h_F}{h_D}T_F\right) + h_F T_F - J\Delta H}{k_m + h_F\left(1 + \frac{k_m}{h_D}\right)} \tag{6}$$

$$T_2 = \frac{k_m\left(T_F + \frac{h_D}{h_F}T_D\right) + h_D T_D + J\Delta H}{k_m + h_D\left(1 + \frac{k_m}{h_F}\right)} \tag{7}$$

where k_m is the heat transfer coefficient through the membrane [W/mK], h is the convective heat transfer coefficient [W/m^{2}K], and ΔH is vapour enthalpy [kJ/kg]. The coefficient h was calculated from equation:

$$h = \frac{Nu\ k_L}{d_h} \tag{8}$$

where Nu is the Nusselt number, k_L—liquid (feed or distillate) heat transfer coefficient, and d_h—hydraulic diameter of module channel. The value of Nu number was calculated from correlation [46]:

$$Nu = 4.36 + \frac{0.036Pe\frac{d_h}{L}}{1 + 0.0011\left(Pe\frac{d_h}{L}\right)^{0.8}} \tag{9}$$

where Pe is the Peclet number, L—length of channel (e.g., membrane capillary).

The above equations show that the temperatures T_1 and T_2 significantly depend on the flow turbulence; hence, in addition to the T_F and T_D values the MD process performance is also strongly affected by the flow velocity of the streams and the dimensions of the channels in a module.

A temperature difference ($T_F - T_D$) causes the distillate temperature to increase not only as a result of vapour condensation, but also due to the heat conduction through the membrane from feed to distillate stream. Therefore, a temperature of the boundary layer (T_2) rises quickly, and the temperature in the distillate channel (T_D) becomes close to the feed temperature (Figure 2). For this reason, it is important to determine the useful length of the MD module, which will depend not only on the flow velocity of the streams and their initial temperature [29,34,47], but also on the diameter of the capillary membranes.

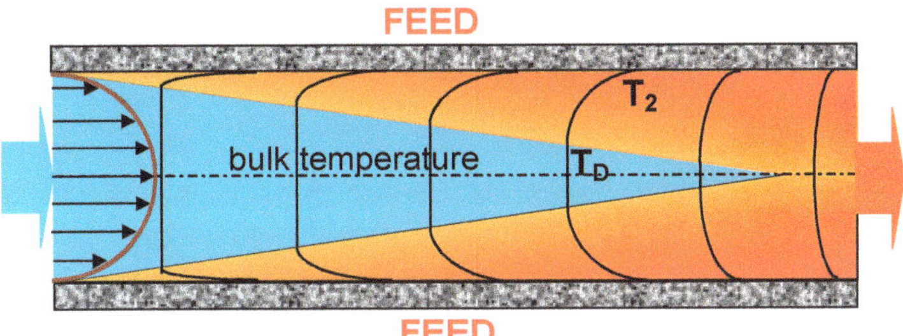

Figure 2. Scheme of distillate temperature changes during distillate flow inside the capillary membrane. T_D—distillate temperature.

The changes of the distillate and feed temperatures along the capillary membranes were calculated based on the balance equations formulated for a differential element (ΔL) presented in Figure 1:

$$\left(v_{F[i]}A_F\rho_F - J_{[i]} \cdot dF\right) Cp_F\left(T_{F[i]} - T_{F[i+1]}\right) = h_F dF\left(T_{F[i]} - T_{1[i]}\right) \quad (10)$$

$$\left(v_{D[i]}A_D\rho_D + J_{[i]} \cdot dF\right) Cp_D\left(T_{D[i+1]} - T_{D[i]}\right) = h_D dF\left(T_{2[i]} - T_{D[i]}\right) \quad (11)$$

where ρ is the liquid density, dF is the membrane area for capillary length of ΔL, and A is the cross-section area of channel [m^2] for the feed and distillate sides, respectively.

The presented MD model was experimentally validated in our previous works [44,46]. The numerical simulations were realized using distilled water as a feed, applying the physical-chemical parameters (such as C_P, ρ, ΔH, and k_L) published in work [48].

3. Materials and Methods

The submerged modules were tested using the DCMD variant in the installation schematically shown in Figure 3. In most cases, the membrane module was located in the feed tank and the distillate flowed inside the capillary membranes. For comparison, the tests were also carried out in which the module was immersed in a distillate tank and the feed flowed inside the capillaries. The volume of feed and distillate tanks were 4 and 1 L, respectively. The maximum efficiency of the peristaltic pump (type 372A, Elpin-Plus, Lubawa, Poland) was 0.5 L/min, which allowed examination of the submerged modules with the flow rate in the range of 0.2–1.4 m/s (depending on module construction). The hydrostatic pressure induced by the water flow was measured at the module inlet using a pressure gauge (KMF, Wika, Włocławek, Poland) with an accuracy of ±100 Pa.

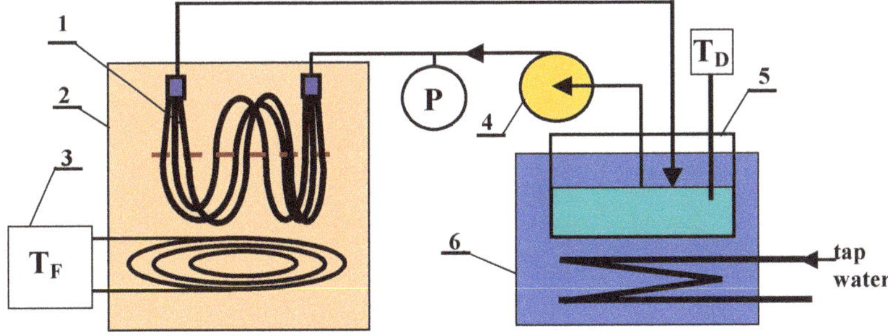

Figure 3. MD process experimental set-up. 1—submerged MD module, 2—feed tank, 3—feed temperature regulator with electric heater, 4—peristaltic pump, 5—distillate tank, 6—cooling bath, T—thermometer, P—pressure gauge.

In the conducted MD studies, the distillate temperature was 293 ± 1 K and the feed temperature was varied in the range of 323–358 K. An electric heater with electronic temperature regulator (± 0.5 K) was applied (type EDIG, Nüga Company, Georgensgmünd, Germany). The electronic thermometers (PT 401, Elmetron, Poland) were used to measure the temperatures with an accuracy of ±0.1 K. Distilled water with the addition of NaCl (1 g/L) was used as the feed, which allowed checking of feed leakage. The value of electrical conductivity (6P Ultrameter, Myron L Company, Carlsbad, CA, USA) of water on the distillate side did not exceed 10 μS/cm in each of the examined cases, which proved that the module construction was tight.

Most MD tests were performed without mechanical mixing of the feed, and the feed flow between the capillaries was caused only by the natural convection generated by the electrical heater assembled below the submerged module (Figure 3). Moreover, to demonstrate the effect of temperature polarization on the external side of the capillaries, the additional experiments were carried out with mixing of water in the feed tank (No. 2 in Figure 3) by the magnetic stirrer (700 rpm, RCT type, IKA-Werke GmbH, Staufen, Germany).

The commercial capillary membranes fabricated for the microfiltration process (MF) were used to construct the submerged modules. The capillaries were made from different polymers, and membranes exhibited the significant differences in their parameters (Table 1). These membranes were assembled in modules of various constructions that are schematically shown in Figure 4. The parameters of tested modules are presented in Table 2.

Table 1. Commercial capillary membranes assembled in submerged MD modules. Membrane parameters—manufacturers' data.

Manufacturer	Membrane	Polymer	d_p [μm]	Porosity [%]	D_{in} [mm]	Wall Thickness [mm]
EuroSep Poland	EuroSep PP	PP	0.20	70	1.8	0.4
PolyMem Poland	K1800	PP	0.20	74	1.8	0.35
Membrana Germany	Accurel PP S6/2	PP	0.22	73	1.8	0.4
Membrana Germany	Accurel PP V8/2 HF	PP	0.20	73	5.5	1.5
Memtek, Australia	HL310	ECTFE	0.40	75	0.31	0.17
Memtek, Australia	PV370	PVDF	0.10	74	0.37	0.12

d_p—pore diameter, D_{in}—capillary membrane inner diameter.

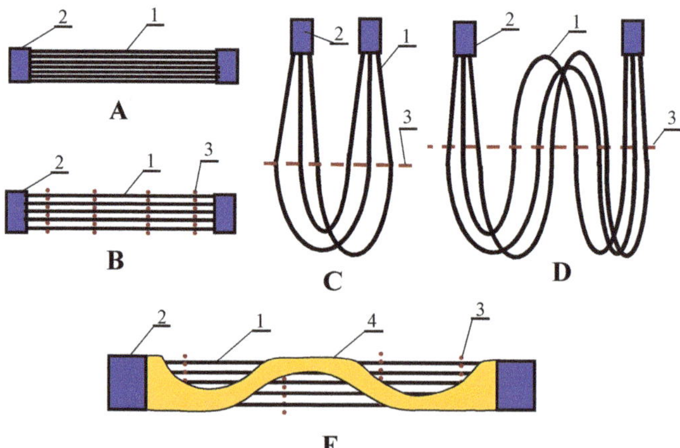

Figure 4. Various constructions of submerged MD module. (**A**) statistical arrangement of capillaries; (**B**) capillaries located inside the nets; (**C**) U-shape type, membranes located inside the net; (**D**) multiple U-shape; (**E**) variant 3B stiffened by tubular housing with cut out large holes. 1—capillary membrane, 2—module head, 3—PP grid, 4—tubular housing with holes.

Table 2. Parameters of submerged MD modules.

Module	Membrane	Number of Capillaries	Length [cm]	Area (Inside) [cm^2]	Mode (Figure 4)
SMD1	EuroSep PP	3	85.5	145.0	D
SMD2	K1800	3	84.7	143.7	D
SMD3	Accurel PP S6/2	2	60.5	68.4	D
SMD4	PV370	19	120	265.3	D
SMD5	Accurel PP S6/2	4	22.5	50.9	C
SMD6	Accurel PP V8/2 HF	1	26.5	45.8	B
SMD7	HL310	75	13.0	95.1	B
SMD8	PV370	85	10.1	73.9	B
SMD9	K1800	3	20.0	98.8	B
SMD10	PV370	55	7.5	47.9	E
SMD11	PV370	55	8.7	55.6	B
SMD12	Accurel PP S6/2	2	120	136.7	D
SMD13	Accurel PP V8/2 HF	1	69	119.5	B

To compare the performance of submerged modules with the conventional modules, additional MD tests were performed using the capillary modules (Table 3). The membranes inside the modules with housing were assembled in grids, i.e., similarly to the modules shown in Figure 4B,E. This mode provided good hydrodynamic conditions for liquid flow along the capillaries, which allowed significant increase of the efficiency of capillary modules [49].

Table 3. Parameters of capillary MD modules.

Module	Membrane	Number of Capillaries	Length [cm]	Area (Inside) [cm^2]	Housing Diameter [mm]
CMD1	Accurel PP S6/2	8	65	294	12
CMD2	Accurel PP V8/2 HF	1	68	117	12

4. Results and Discussion

4.1. *Influence of MD Process Parameters*

In the case of submerged modules there are two options of MD process operation. In the first option (Mode 1), the modules are submerged in a feed tank and the distillate flows inside the capillaries. In Mode 2 the configuration is reversed, i.e., the modules are submerged in the distillate tank. In each of these modes the possibility of enhancement in the flow turbulence mainly exists in the lumen of the capillaries. In this work, the majority of studies were carried out using Mode 1, in which the modules were immersed in the hot feed. Moreover, the additional series of measurements was also performed with Mode 2 for comparison purposes.

An advantage of submerged MD modules is maintaining a constant temperature outside the membranes, but the temperature of stream flowing in the lumen undergoes the changes (feed cools down or distillate heats up), which decreases the process efficiency. With regard to an exponential dependence of vapour pressure on a temperature, it can be expected that the changes of feed temperature will have a larger influence on a decline of process efficiency. The results presented in Figure 5 confirmed that the permeate flux was almost two times higher when the submerged MD module was placed in the feed tank (Mode 1) compared to the distillate tank (Mode 2). In the former case the entire surface of the capillaries was in contact with the feed at the constant high temperature (e.g., 353 K). In the latter option (Mode 2), the feed flowed inside the capillaries and the feed temperature systematically decreased along the module length due to the mass and heat transfer. As a result, the outlet feed temperature (T_{Fout}) was lower from its inlet value (T_{Fin}) and a decline of the permeate flux was observed (Figure 5).

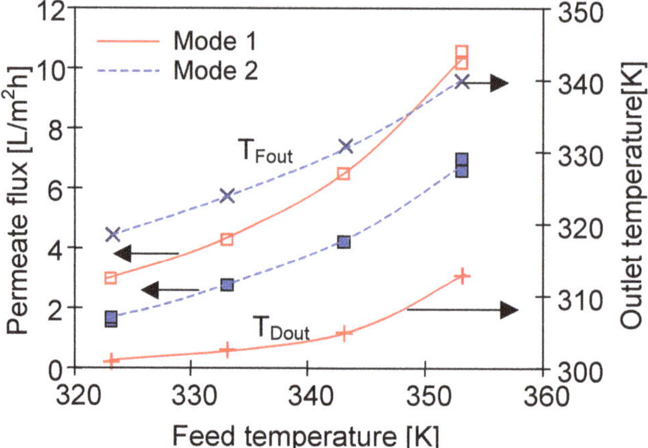

Figure 5. The influence of feed temperature and direction of feed flow (Mode 1 or Mode 2) on the permeate flux and outlet temperature of feed or distillate. Module SMD1. Flow rate inside the capillaries v = 0.62 m/s. Distillate inlet temperature 293 K.

A decrease in the feed temperature (Mode 2) can be also limited by increasing the flow rate of the feed, which allows the permeate flux to increase, especially for a higher inlet feed temperature (Figure 6). However, the studies confirmed that even for high flow rates, it was impossible to prevent the temperature decline of the feed flowing inside the capillaries. In this case, a significant part of the advantages associated with the increase in the flow rate resulted from the increasing turbulence of flow, which limited the negative effect of temperature polarisation [29,40,47,50]. Since the hydrostatic pressure increased with the increase in the flow rate, the possibility of membrane wetting was also enhanced in this case.

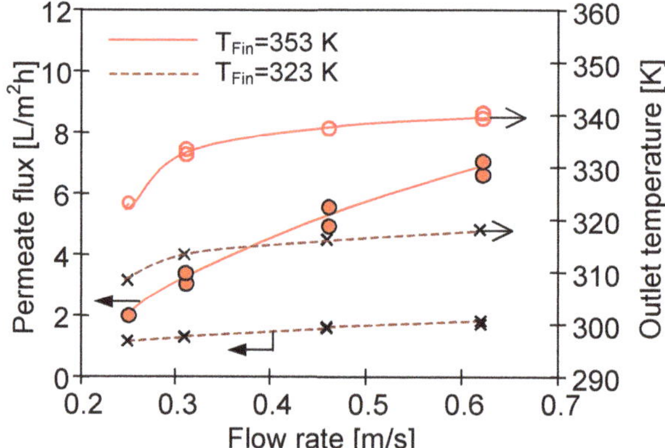

Figure 6. The influence of feed flow velocity on the permeate flux and feed outlet temperature for two different inlet temperatures (T_{Fin}). Mode 2—distillate temperature 293 K, feed inlet temperature 323 K or 353 K. Module SMD1.

The above-presented results confirmed that a placement of submerged modules in the feed tank was more advantageous. In this configuration (Mode 1) the distillate flowed inside the capillaries, and the distillate temperature gradually increased, moreover, such an option reduced the process efficiency to a smaller degree (Figure 5). This resulted from the character of changes in the driving force, which in the DCMD was the vapour pressure difference (Δp) between the feed and distillate [39,51,52]. The water vapour pressure increased exponentially, exhibiting a significant growth just only above 333 K [39,43,50]. Therefore, the increase in the distillate temperature, even from 303 to 333 K (ΔT = 30 K, Δp = 15.6 kPa), did not cause such a large decrease of the driving force as the decline of feed temperature from 353 to 323 K (ΔT = 30 K, Δp = 35 kPa). Therefore, the MD efficiency obtained for Mode 1 (e.g., T_F = 343 K) was similar to that obtained for higher feed temperature (T_{Fin} = 353 K, v_F = 0.62 m/s) with the feed flowing inside the capillaries (Figure 5, Mode 2).

In an analogous manner, the increase of the distillate flow rate can limit an elevation of distillate temperature inside the capillaries. Due to the abovementioned dependence of the vapour pressure on temperature, a slight growth of the permeate flux (as a function of flow rate) was obtained, mainly for the distillate outlet temperature in the range of 313–327 K (Figure 7). On that account, it was advantageous to select such a flow rate of distillate that allowed a distillate outlet temperature below 313 K. In the case of the tested module SMD2 such flow rate was above 0.4 m/s (Figure 7).

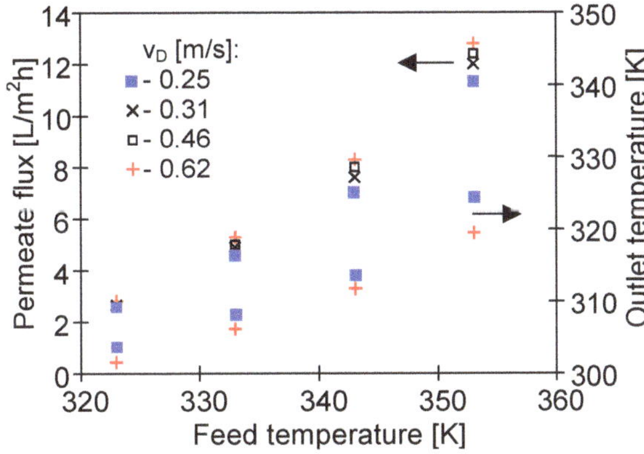

Figure 7. The influence of feed temperature and the distillate flow rate on the permeate flux and distillate outlet temperature. T_{Din} = 293 K. Mode 1. Module SMD2.

The numerical modelling revealed that the negative effect of the increase of the distillate temperature would be enhanced with the increasing length of capillary membranes. The simulation results for Accurel PP S6/2 membranes with a length of 120 cm are presented in Figure 8. An extension of the module length from 80 to 120 cm caused the temperature T_{Dout} to increase, e.g., for v_D = 0.6 m/s from 309.5 to 316.6 K. The calculation results also confirmed the significant effect of the flow rate on the value of temperature polarization. The difference between T_2 and T_D was reduced more than twice by increasing the velocity v_D from 0.2 to 0.8 m/s. The results of experimental studies confirmed the conclusions resulting from the numerical simulations. The performance of the SMD12 module shown in Figure 9 (length 120 cm) proved a greater influence of distillate flow rate changes than that observed for a shorter SMD2 module (Figure 7).

In the MD process the feed is a source of energy, hence, for the continuous separation it is necessary to ensure a minimum feed flow along the fibres of the submerged modules. In the calculations presented in Figure 8 it was assumed that the feed flows at a velocity of 0.05 m/s along the membranes (e.g., due to the air bubbling or feed recirculation) and that a distance between the capillaries is 7 mm. A significant influence of turbulence of the feed flow on the MD performance was confirmed by the results presented in Figure 9. Activation of a magnetic stirrer in the feed tank caused an increase of the SMD12 module efficiency by over 20%. Although the bulk temperature of feed along the entire surface of the membranes was similar, the temperature T_1 was lower than T_F, due to the water evaporation and heat conduction from the feed to the distillate (Figure 1).

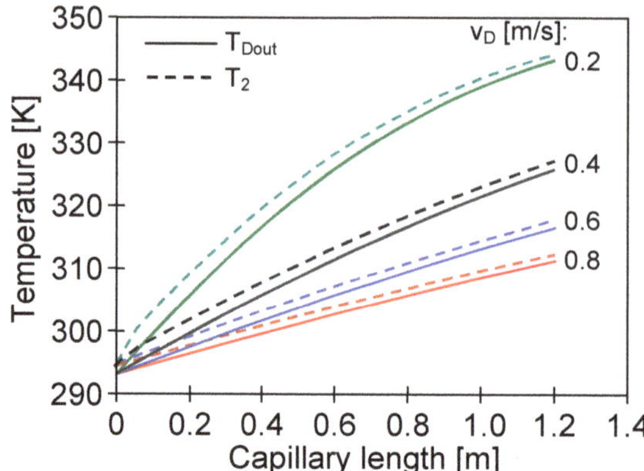

Figure 8. The results of numerical calculations. The influence of distillate flow rate on distillate bulk temperature and temperature of condensation layer (T_2). Mode 1: T_F = 353 K, T_{Din} = 293 K. Membranes: Accurel PP S6/2, L = 120 cm. Feed flow v_F = 0.05 m/s.

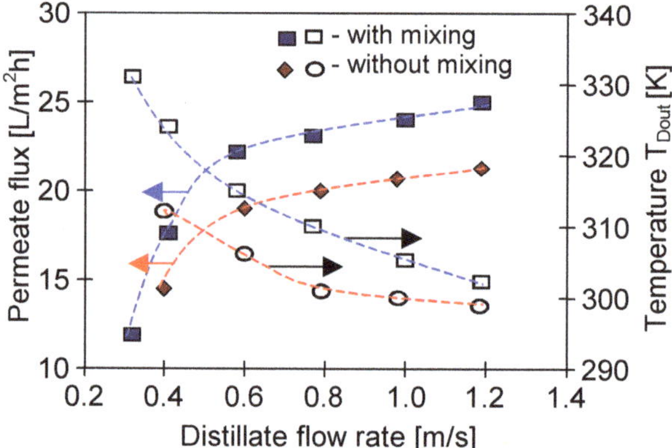

Figure 9. The influence of distillate flow rate and turbulence in the feed tank on the permeate flux and distillate outlet temperature in the presence and absence of feed mixing. Mode 1: T_F = 353 K. Module SMD12 (Accurel PP S6/2, L = 120 cm).

4.2. The Effect of Membrane Morphology

The membrane parameters, such as a pore size and wall thickness, will influence the quantity of energy transferred through the membrane [25,35]. Therefore, the temperature in the lumen of the capillary varies differently depending on the membrane type, which has an influence on the module yield as it was demonstrated. On that account, in this part of the study shorter modules (SMD5-9) were employed, allowing an outlet temperature closer to the inlet values. The obtained results of the studies are presented in Figure 10. In each tested case, the permeate flux was stable over the time of module operation. However, the flux values differed significantly for the particular modules. One of the main reasons was the different thickness of the membrane walls and various pore sizes. The highest efficiency (19 L/m^{2}h) was achieved for the SMD7 module with the ECTFE membranes having the wall thickness of 170 μm and 0.4 μm pore size, whereas the lowest permeate flux (6.7 L/m^{2}h) was obtained for PP membranes with the wall thickness of 1500 μm and 0.2 μm pore size. The results obtained for the membranes with similar capillary diameters and wall thickness (Table 1, SMD5 and SMD9) indicated that a slightly larger value of the permeate flux may result from a higher porosity and larger pore size [53,54]. For similar reasons the efficiency of the SMD7 module was higher than that obtained for the SMD8 module with PV370 membranes, which had the smallest pore diameters (0.1 μm) (Table 1).

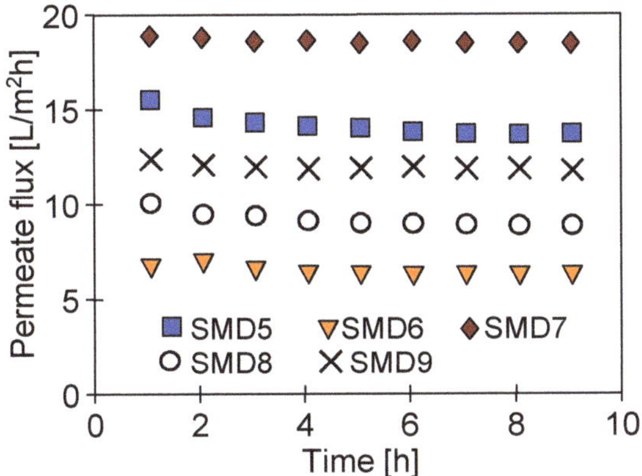

Figure 10. The influence of membrane type on the permeate flux. Mode 1: T_F = 353 K. Membranes: SMD5—Accurel PP S6/2; SMD6—Accurel PP V8/2 HF; SMD7—HL310; SMD8—PV370; and SMD9—K1800.

The influence of membrane morphology and the operating conditions of the MD process on the efficiency of the modules presented in this work have been also reported by different authors in the studies of submerged MD configuration [13,15,24,30–33,55,56]. The results compiled in Table 4, especially a comparison of the permeate flux values, indicate that due to the limited turbulence in the feed flow, the efficiency of the submerged modules was at average level. For example, the permeate flux in the range of 4–13 L/m^{2}h should be expected for submerged modules when the feed temperature is 343 K. Moreover, the experimental results presented above indicate that obtained performance can be increased if the proper hydrodynamic conditions are applied in the MD installation.

Table 4. Comparison of the results of the MD process obtained for different submerged MD modules.

Process	Membrane	D_{in} [mm]	Wall Thickness [μm]	T_F [K]	TDS [g/L]	Flux [L/m^{2}h]	Ref.
VMD	Accurel PP S6/2	1.8	400	343	14.2	8.2	[15]
VMD	Accurel PP S6/2	1.8	400	343	Demi 200	9.0 6.0	[30]
VMD	Accurel PP S6/2	1.8	400	343	6.92	13.0	[55]
VMD	PTFE	0.86	415	348	1.8	4.2	[31]
VMD	PTFE	0.8	450	348	100	6.23	[32]
VMD	PVDF	0.7	250	328	demi	14.1	[13]
DCMD	Accurel PP S6/2	1.8	400	343	Demi 200	4.6 3.0	[30]
DCMD	Accurel PP S6/2	1.8	400	-	demi	4.6	[56]
DCMD	Accurel PP S6/2	1.8	400	-	-	8.2	[24]
DCMD	PVDF	0.7	250	328	demi	8.5	[13]
DCMD	PVDF	0.7	250	328	12	2.7	[57]
DCMD	PVDF	0.7	250	328	demi	7.9	[13]

4.3. The Effect of Capillary Membrane Length

Not only the structure of a membrane wall but also the diameter of a capillary has a significant influence on the performance of the MD submerged modules. In the case of membranes with a diameter of D_{in} = 1.8 mm, the distillate outlet temperature was much lower than the feed temperature after passing a 120 cm long capillary, even at lower distillate flow rates (Figures 7–9). However, for capillaries with a smaller diameter, the temperature of distillate flowing inside the capillaries can approach the feed temperature (Figure 2), which will result in a large decrease in module efficiency. This dependence was confirmed by the results of the MD studies with SMD4 module (PV370 membrane, D_{in} = 0.37 mm) presented in Figure 11, for which the obtained permeate flux values were twice lower than those obtained for the SMD12 module with capillary membranes having the same length. Moreover, the permeate flux increased by over 25% when the magnetic stirrer in the feed tank was running, which indicated that the temperature polarization on the feed side had a large impact on the performance of the SMD4 module, similar to the case of the SMD12 module (Figure 9). A significant influence of feed agitation on the submerged modules performance was also presented in work [26].

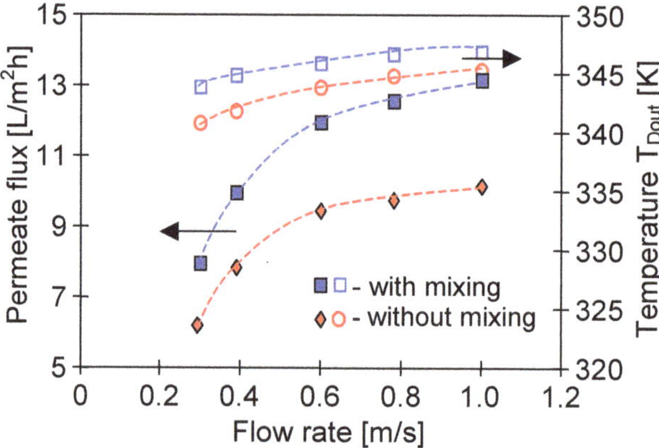

Figure 11. The influence of distillate flow rate and turbulence in the feed tank on the permeate flux and distillate outlet temperature in the presence and absence of feed mixing. Mode 1: T_F = 353 K. Module SMD4 (PV370, L = 120 cm).

The numerical analysis of the MD process with PV370 membranes revealed that the distillate temperature inside capillaries was close to the feed temperature after flowing through a distance of 20–30 cm, even for high flow velocities (e.g., 1.2 m/s) (Figure 12, T_F = 353 K). Due to such a significant increase in the distillate temperature, more than half of the module working area was practically inactive (Figure 13). A large part of the membrane area was also inactive when the amount of energy transferred to the distillate was substantially reduced by decreasing the feed temperature to 313 K (Figure 12). Moreover, the distillate temperature was lower than 330 K in each of the analysed cases (T_F = 353 K) for capillaries having up to 20 cm length, which allowed the generation of the appropriate driving force for vapour transfer. For this reason, a good MD process performance can be obtained by using short (e.g., 10–20 cm) modules in laboratory tests. However, these results cannot be reproduced in the industrial installations with longer modules. A similar conclusion was also drawn by other authors [26].

Although an increase of the flow rate is an effective way of lowering the distillate temperature, this option simultaneously causes an increase of the hydrostatic pressure. In the case of PV370 membranes, for v_D = 1.2 m/s this pressure was increased to 0.14–0.16 MPa (module SMD4), which created a risk of exceeding of the Liquid Entry Pressure (LEP) value. Nevertheless, even at 0.18 MPa no leakage was observed. These results indicated that the obtained hydrostatic pressure was lower than LEP. However, it is worth mentioning that each increase of pressure accelerated a phenomenon of membrane wetting during the long-term module exploitation. Therefore, also for this reason, the capillary membranes with the diameters definitely larger than the tested PVDF membranes (D_{in} = 0.37 mm) should be applied in the submerged modules.

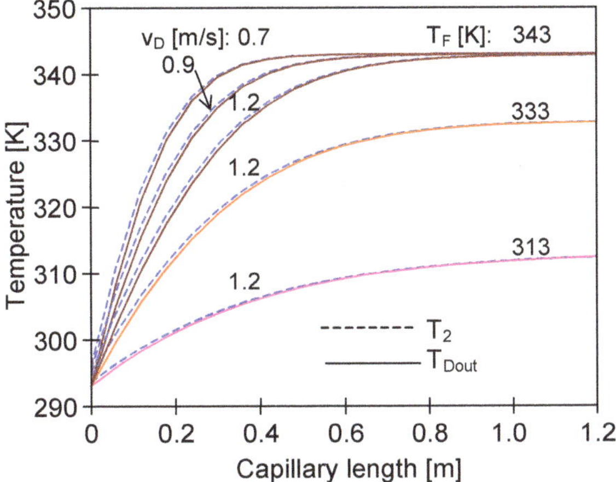

Figure 12. The influence of distillate flow rate and T_F on the distribution of the T_2 and distillate outlet temperature along the membrane length. Mode 1. Membranes PV370, L = 120 cm.

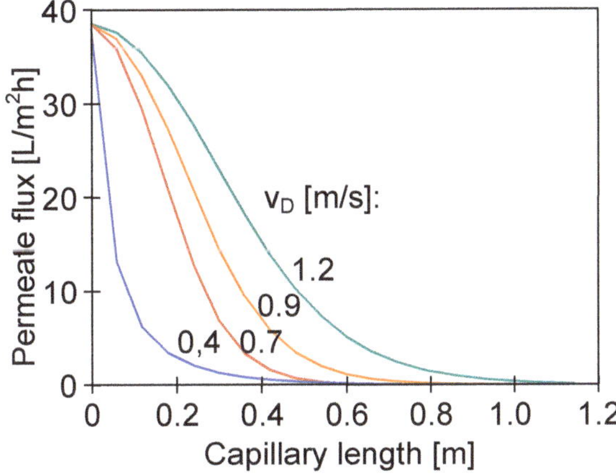

Figure 13. The influence of distillate flow rate on the distribution of the permeate flux along the module length. Mode 1. T_F = 343 K. Membranes PV370.

The presented results indicate that determination of the appropriate diameter of the capillary membranes that can be used in the submerged MD modules depends on the process temperature and a required length of the module. Figure 14 shows the calculation results obtained for membranes with a diameter in the range of 0.37–1.0 mm. For the analyzed MD process conditions and the capillary membrane diameter equal to 1.0 mm, the obtained T_{Dout} value was about 10 K lower than T_F with a membrane length of 0.8–1 m. Therefore, the minimum diameter should be 1 mm for capillary membranes assembled in the industrial modules.

The latter conclusion was confirmed by tests using the SMD3 module, in which 1.8-mm diameter membranes were used. In this case, the hydrostatic pressure increased only to 0.032 MPa, although the flow rate was increased to almost 1.4 m/s (Figure 15). Moreover, the application of the membranes with a larger diameter (over 1 mm) also allowed lower distillate temperature with the increase of flow rate. In the discussed experiments the T_{Dout} value approached the inlet value for a flow rate of over 0.6 m/s.

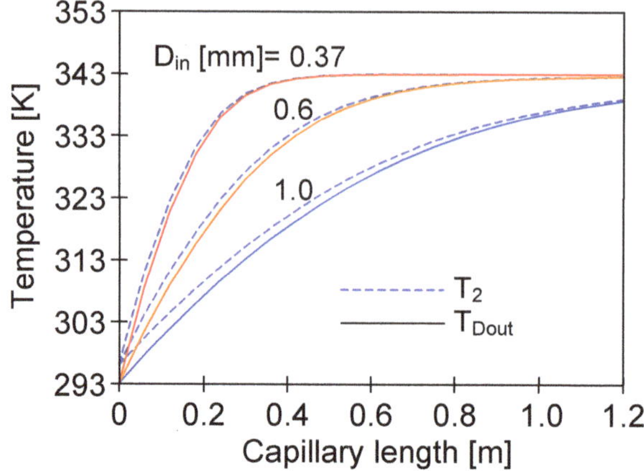

Figure 14. The influence of capillary diameter on the feed outlet temperature. Mode 1. Membrane PV370, T_{Din} = 293 K, T_F = 343 K, v_D = 0.7 m/s and v_F = 0.05 m/s.

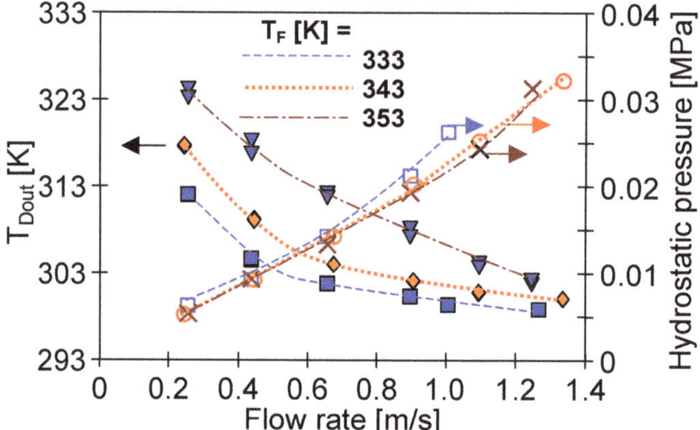

Figure 15. The influence of distillate flow rate on distillate outlet temperature and hydrostatic pressure inside capillaries. Module SMD3 (Membrane Accurel PP S6/2). Mode 1: T_F = 353 K.

4.4. Module Construction

The results presented in Figures 5–7 were obtained using U-shape modules with the membranes assembled in the spacing grids (Figure 4D). Such a solution achieved a distance of 3–10 mm between the membranes. Such significant distances permitted a free flow of the feed between the capillaries, hence, the effect of stream distributions on the process efficiency was eliminated. However, these modules had a small packing density, hence, a construction in which the membranes were assembled in the fibre bundles (e.g., Figure 4A), allowing definitely larger membrane areas calculated per unit volume of the feed tank. A basic issue will be the prevention from excessive decline of the feed temperature inside such a bundle due to a high resistance of the feed flow between the capillaries. Therefore, one of the important construction parameters is the distance in which the fibres should be placed from each other. The results obtained for the modules SMD7 and SMD8 are presented in Figure 10. In the SMD8 module the membranes were arranged at the distance of 0–1 mm, whereas in the SMD7 module the spacing grids were applied, and as a result, the distance between the capillaries amounted to 2–3 mm. The resulting lower permeate flux for the SMD8 module confirmed that a too-high packing density (PD) of fibres decreases the efficiency of the submerged MD module. A value of PD is usually expressed by a ratio of the capillary volume to the tank volume [26], which can be calculated from the following equation:

$$PD = \frac{N\,D_{out}^2}{D_V^2} \tag{12}$$

where N—a number of capillaries, D_{out}—external diameter of capillary membrane, D_V—vessel diameter. The equation is valid when the capillaries are uniformly distributed in the tank.

The results obtained from the numerical analysis, which takes into account the effect of membrane packing density on the submerged module performance, are shown in Figure 16. In these calculations it was assumed that the capillary membranes were vertically submerged in a cylindrical vessel with exemplary diameter equal to 5 cm. Moreover, the calculations were performed with the assumption that the feed flowed from the bottom of the tank along the membrane bundle (120 cm) at a velocity of 0.05 or 0.1 m/s, and with an initial temperature T_{Fin} = 353 K. In both cases, the flow rate of distillate was v_D = 0.4 m/s. The number of capillaries in the tank was increased from 20 to 250, which resulted in a change in packing density values from 5% to 68%.

An increase of the packing density caused a deterioration of the module operating conditions, although the MD process efficiency increased more than twice for v_F = 0.1 m/s. It was found that even such a small increase in the flow rate hindered the reduction of the feed temperature T_F, as well as improved the conditions of water evaporation (increase of T_1). The obtained results confirmed the conclusions of the previous experimental studies, which indicated that the membrane packing density at a level of 40% was advantageous for MD modules [49]. In the case of arrangement of the capillary membranes in the orthogonal system, the considered Accurel PP S6/2 membranes will be at a distance of about 2 mm from each other, if 40% packing density is applied.

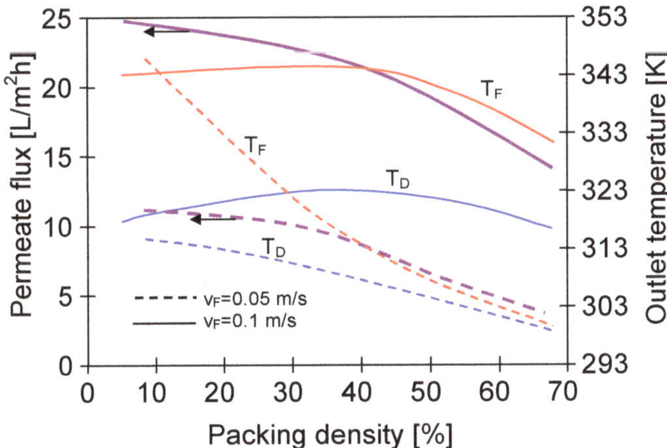

Figure 16. The influence of membrane packing density on the permeate flux and feed outlet temperature for various feed flow values. Mode 1. Distillate inlet temperature 293 K, feed temperature 353 K. v_D = 0.4 m/s. Membrane Accurel PP S6/2.

4.5. Comparison of Submerged and Capillary MD Modules

The above results clearly indicate that both the diameter and length of the capillary membranes assembled in the MD submerged modules, as well as their packing density, were the important construction parameters. Maintaining the appropriate construction and operating conditions (feed turbulence and larger capillary diameter), it was possible to achieve an almost constant feed temperature outside the submerged modules. However, the temperature of distillate flowing inside the capillary membranes increased with the increase of module length. If the temperature of the distillate exceeded 320 K, this caused a significant decrease in the module efficiency (e.g., Figure 9, v_D = 0.3 and 0.4 m/s). In the conventional cross-flow MD module configuration with housing, the feed temperature decreased along the module; hence, the elevation of the distillate temperature in this module was smaller than that in the submerged modules (Figure 17). However, the temperature increase in the range of 315–320 K did not cause such a significant increase in the water vapour pressure. Therefore, when the distillate temperature was maintained at this level, the efficiency of the submerged module similar to that obtained in the conventional capillary modules can be achieved (Figures 9 and 17). Heat conduction from the feed to the distillate (heat loss) can be reduced by using thick-walled membranes; however, in this case the mass transfer resistance was also increased, resulting in a decline in the process efficiency (Figure 18). The performance of the SMD13 submerged module in this case was similar to the CMD2 capillary module.

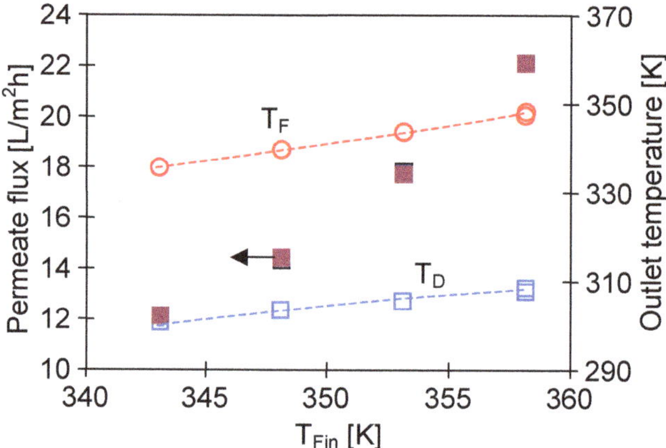

Figure 17. The influence of feed temperature on the permeate flux and feed/distillate outlet temperature. Capillary module CMD1. T_{Din} = 290 K, v_F = 0.8 m/s, v_D = 0.4 m/s.

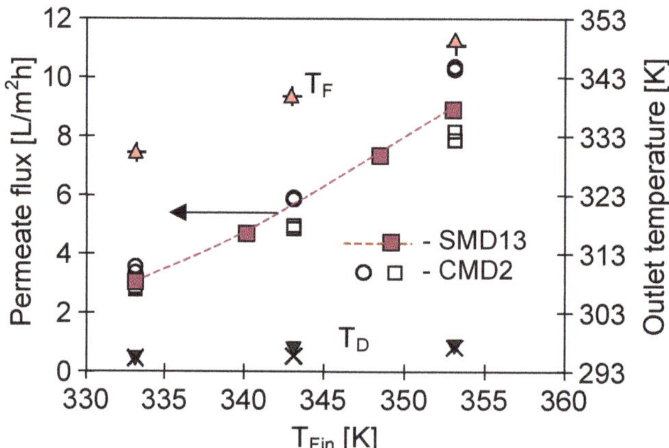

Figure 18. The influence of feed temperature on the permeate flux and feed/distillate outlet temperature. Capillary module CMD2. T_{Din} = 290 K, v_F = 0.6 m/s, v_D = 0.3 m/s. □▲▼—feed on the housing side; ○+x—feed inside the capillary. ■—flux obtained for SMD13 module (Mode 1).

The efficiency of the MD process increased with the decrease of membrane wall thickness. Unfortunately, the mechanical strength of capillaries was reduced in this case. The membranes in the submerged modules were subjected to a significant stress, hence, reinforced membranes were used in the MF or UF industrial modules [1]. In the case of MD, one solution can be assembling the capillary membranes in supporting constructions, eliminating an excessive stress, particularly at a bonding layer in the module head. For example, a tubular housing as in the typical capillary module can be used. However, the housing will possess the holes, which enable a free flow of the feed from a tank to the space between the fibres (Figure 4E). In the examined case the SMD10 module was applied, in the housing of which three large holes were cut out, leading to a 40% loss of the housing area. The obtained MD permeate fluxes were compared with the results for the SMD11 module with the same membranes but without the external housing. A high and stable process efficiency was achieved in both cases (Figure 12). A slightly higher permeate flux obtained for the SMD10 module resulted most probably from the differences in the fibre arrangement in a module and from the fact that the fibres were shorter by about 15%. Moreover, such a result indicated that the application of the housing with the large holes did not deteriorate the MD process conditions.

In the MD process a phase transition (water evaporation) takes place, hence, the process requires delivery of a large quantity of energy. For this reason, the thermal efficiency of the process is important, and it can be determined from the following relationship:

$$E = Q_D/Q_T \tag{13}$$

where Q_D is the latent heat of distillate and Q_T is the total heat transferred from feed to the distillate stream.

The calculated values of the thermal efficiency for submerged modules did not exceed 40% (Figure 19), and these values were significantly lower than those obtained for MD capillary modules with housing, which achieved the efficiency at a level of 70–80% [58]. In the performed studies (Figure 19), mixing of feed was not used, and the convective currents caused by heaters located under the modules only forced the feed flow. For this reason, it can be expected that a better efficiency of submerged modules is possible under enhanced flow turbulence on the feed side. One of the solutions is air bubbling applied in the industrial submerged MF or UF modules [1,30,38]. Aeration is also advantageous in the MD process [22,23], but one should remember that in certain cases the air bubbles can also initialize crystallization, which contributes to the membrane scaling [15,33]. Another option is the application of the mechanical mixing. In the case of the SMD12 module (Figure 9, v_D = 1–1.2 m/s), running the magnetic stirrer increased the thermal efficiency from 40% to 65%.

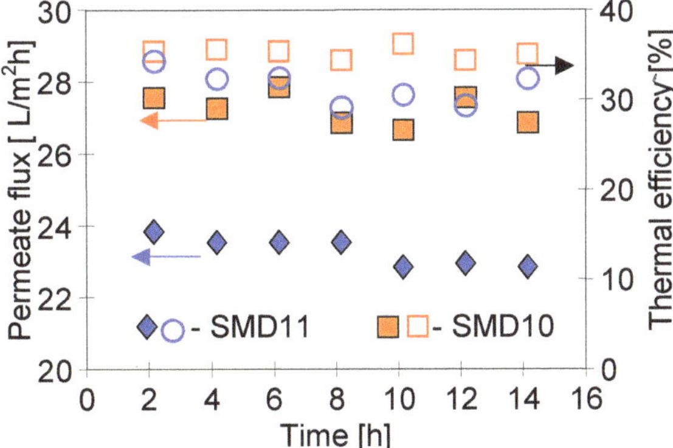

Figure 19. The influence of module construction on the permeate flux and thermal efficiency. Membrane PV370. Module design: SMD10—Figure 4E, SMD11—Figure 4B. Mode 1. T_F = 353 K, v = 0.75 m/s.

5. Conclusions

In the recent years, the interest in membrane distillation has significantly increased and several hundred of articles on MD were published. Nevertheless, only a few of them raise the issue of submerged modules, which can be an interesting alternative to the classical capillary modules with the housing. The results of the research described in the present paper demonstrate that a high MD process efficiency similar to the capillary modules can be also achieved using the submerged modules.

A placement of MD submerged modules in the feed tank allowed a constant feed temperature along the entire membrane surface. Hence, in that case the efficiency of the MD process was twice as high as in the configuration in which the feed flowed inside the capillaries and the modules were submerged in a cold distillate. A significant increase in the module efficiency can be achieved for the feed flowing between the fibres at a velocity of at least 0.05–0.1 m/s.

The application of high packing density of the membranes in the module will hinder the flow of feed between the capillaries, and as a consequence it can lead to a local drop of the feed temperature. The obtained results indicate that the packing density at a level of 40% is advantageous for submerged modules. Under these conditions, for tested capillary membranes (external diameter 2.6 mm) the distance between particular capillaries should amount to about 2–3 mm.

A high feed temperature along the entire module surface causes a significant increase of the distillate temperature. When T_D value reaches 320–330 K, a significant decrease of the process efficiency is observed. Using the membranes with a diameter of at least 1 mm, the T_D increase can be limited by increasing the flow rate of the distillate over 0.4 m/s. For membranes with a diameter below 1 mm, the T_D becomes close to the feed temperature after the distillate flows through a 20–30 cm distance along the capillary, even for the high flow rates (1.2–1.4 m/s). In this case, more than half of the module area (for L = 120 cm) is not working and the permeate flux is close to zero in this zone.

In the case of the membranes with a small diameter an increase of distillate flow rate causes a significant growth of hydrostatic pressure, which can also accelerate their wetting. For example, for the membrane with a diameter of 0.37 mm, an increase of the flow rate from 0.4 to 0.78 m/s caused an increase of hydrostatic pressure from 65 to 155 kPa. The application of the capillary membranes with larger diameters allows definitely smaller values of pressure, e.g., for the tested membrane Accurel PP S6/2 with the diameter of 1.8 mm, the hydrostatic pressure was 33 kPa when flow rate was equal to 1.34 m/s.

Funding: This research was funded by National Science Centre, Poland, grant number 2018/29/B/ST8/00942

Conflicts of Interest: The author declares no conflict of interest. The funders had no role in the design of the study; in the collection, analyses, or interpretation of data; in the writing of the manuscript, or in the decision to publish the results.

References

1. Fane, A.G. Submerged membranes. In *Advanced Membrane Technology and Applications*; Li, N.N., Fane, A.G., Ho, W.S.W., Matsuura, T., Eds.; John Wiley & Sons: Hoboken, NJ, USA, 2008; Chapter 10.
2. Khaing, T.H.; Li, J.F.; Li, Y.Z.; Wai, N.; Wong, F.S. Feasibility study on petrochemical wastewater treatment and reuse using a novel submerged membrane distillation bioreactor. *Sep. Purif. Technol.* **2010**, *74*, 138–143. [CrossRef]
3. Tian, J.Y.; Xu, Y.P.; Chen, Z.L.; Nan, J.; Li, G.B. Air bubbling for alleviating membrane fouling of immersed hollow-fiber membrane for ultrafiltration of river water. *Desalination* **2010**, *260*, 225–230. [CrossRef]
4. Akhondi, E.; Zamani, F.; Tng, K.H.; Leslie, G.; Krantz, W.B.; Fane, A.G.; Chew, J.W. The Performance and Fouling Control of Submerged Hollow Fiber (HF) Systems: A Review. *Appl. Sci.* **2017**, *7*, 765. [CrossRef]
5. Wu, Z.C.; Wang, Z.W.; Huang, S.S.; Mai, S.H.; Yang, C.F.; Wang, X.H.; Zhou, Z. Effects of various factors on critical flux in submerged membrane bioreactors for municipal wastewater treatment. *Sep. Purif. Technol.* **2008**, *62*, 56–63. [CrossRef]
6. Kwon, S.G.; Park, S.W.; Oh, D.K. Increase of Xylitol Productivity by Cell-Recycle Fermentation of Candida tropicalis Using Submerged Membrane Bioreactor. *J. Biosci. Bioeng.* **2006**, *101*, 13–18. [CrossRef] [PubMed]

7. Wang, Z.W.; Ma, J.X.; Tang, C.Y.; Kimura, K.; Wang, Q.Y.; Han, X.M. Membrane cleaning in membrane bioreactors: A review. *J. Membr. Sci.* **2014**, *468*, 276–307. [CrossRef]
8. He, L.; Zhang, J.; Wang, Z.W.; Wu, Z.C. Chemical Cleaning of Membranes in a Long-Term Operated Full-Scale MBR for Restaurant Wastewater Treatment. *Sep. Sci. Technol.* **2011**, *46*, 2481–2488. [CrossRef]
9. Günther, J.; Albasi, C.; Lafforgue, C. Filtration characteristics of hollow fiber microfiltration membranes used in a specific double membrane bioreactor. *Chem. Eng. Process. Process Intensif.* **2009**, *48*, 1255–1263. [CrossRef]
10. Akhondi, E.; Wicaksana, F.; Fane, A.G. Evaluation of fouling deposition, fouling reversibility and energy consumption of submerged hollow fiber membrane systems with periodic backwash. *J. Membr. Sci.* **2014**, *452*, 319–331. [CrossRef]
11. Duong, H.C.; Cooper, P.; Nelemans, B.; Cath, T.Y.; Nghiem, L.D. Optimising thermal efficiency of direct contact membrane distillation by brine recycling for small-scale seawater desalination. *Desalination* **2015**, *374*, 1–9. [CrossRef]
12. Lokare, O.R.; Tavakkoli, S.; Khanna, V.; Vidic, R.D. Importance of feed recirculation for the overall energy consumption in membrane distillation systems. *Desalination* **2018**, *428*, 250–254. [CrossRef]
13. Choi, Y.; Naidu, G.; Jeong, S.; Vigneswaran, S.; Lee, S.; Wang, R.; Fane, A.G. Experimental comparison of submerged membrane distillation configurations for concentrated brine treatment. *Desalination* **2017**, *420*, 54–62. [CrossRef]
14. Choi, Y.; Naidu, G.; Jeong, S.; Lee, S.; Vigneswarana, S. Fractional-submerged membrane distillation crystallizer (F-SMDC) for treatment of high salinity solution. *Desalination* **2018**, *440*, 59–67. [CrossRef]
15. Julian, H.; Meng, S.; Li, H.Y.; Ye, Y.; Chen, V. Effect of operation parameters on the mass transfer and fouling in submerged vacuum membrane distillation crystallization (VMDC) for inland brine water treatment. *J. Membr. Sci.* **2016**, *520*, 679–692. [CrossRef]
16. Wang, P.; Chung, T.S. Recent advances in membrane distillation processes: Membrane development, configuration design and application exploring. *J. Membr. Sci.* **2015**, *474*, 39–56. [CrossRef]
17. Guillen-Burrieza, E.; Mavukkandy, M.O.; Bilad, M.R.; Arafat, H.A. Understanding wetting phenomena in membrane distillation and how operational parameters can affect it. *J. Membr. Sci.* **2016**, *515*, 163–174. [CrossRef]
18. Chamani, H.; Matsuura, T.; Rana, D.; Lan, C.Q. Modeling of pore wetting in vacuum membrane distillation. *J. Membr. Sci.* **2019**, *572*, 332–342. [CrossRef]
19. McGaughey, A.L.; Gustafson, R.D.; Childress, A.E. Effect of long-term operation on membrane surface characteristics and performance in membrane distillation. *J. Membr. Sci.* **2017**, *543*, 143–150. [CrossRef]
20. Rezaei, M.; Warsinger, D.M.; Lienhard, J.H.; Duke, M.C.; Matsuura, T.; Samhaber, W.M. Wetting phenomena in membrane distillation: Mechanisms, reversal, and prevention. *Water Res.* **2018**, *139*, 329–352. [CrossRef]
21. Tomaszewska, M.; Białończyk, L. Production of ethanol from lactose in a bioreactor integrated with membrane distillation. *Desalination* **2013**, *323*, 114–119. [CrossRef]
22. Warsinger, D.M.; Servi, A.; Van Belleghem, S.; Gonzalez, J.; Swaminathan, J.; Kharraz, J.; Chung, H.W.; Arafat, H.A.; Gleason, K.K.; Lienhard, J.H. Combining air recharging and membrane superhydrophobicity for fouling prevention in membrane distillation. *J. Membr. Sci.* **2016**, *505*, 241–252. [CrossRef]
23. Rezaei, M.; Warsinger, D.M.; Lienhard, J.H.; Samhaber, W.M. Wetting prevention in membrane distillation through superhydrophobicity and recharging an air layer on the membrane surface. *J. Membr. Sci.* **2017**, *530*, 42–52. [CrossRef]
24. Zhong, W.W.; Li, H.Y.; Ye, Y.; Chen, V. Evaluation of silica fouling for coal seam gas produced water in a submerged vacuum membrane distillation system. *Desalination* **2016**, *393*, 52–64. [CrossRef]
25. Yazgan-Birgi, P.; Ali, M.I.A.; Arafat, H.A. Comparative performance assessment of flat sheet and hollow fiber DCMD processes using CFD modeling. *Sep. Purif. Technol.* **2019**, *212*, 709–722. [CrossRef]
26. Dong, G.X.; Cha-Umpong, W.; Hou, J.W.; Ji, C.; Chen, V. Open-source industrial-scale module simulation: Paving the way towards the right configuration choice for membrane distillation. *Desalination* **2019**, *464*, 48–62. [CrossRef]
27. Ricci, B.C.; Skibinski, B.; Koch, K.; Mancel, C.; Celestino, C.Q.; Cunha, I.L.C.; Silva, M.R.; Alvim, C.B.; Faria, C.V.; Andrade, L.H.; et al. Critical performance assessment of a submerged hybrid forward osmosis—Membrane distillation system. *Desalination* **2019**, *468*, 114082. [CrossRef]

28. Francis, L.; Ghaffour, N.; Al-Saadi, A.S.; Amy, G.L. Submerged membrane distillation for seawater desalination. *Desalin. Water Treat.* **2015**, *55*, 2741–2746. [CrossRef]
29. Alia, A.; Quist-Jensena, C.A.; Macedonioa, F.; Drioli, E. Optimization of module length for continuous direct contact membrane distillation process. *Chem. Eng. Process. Process Intensif.* **2016**, *110*, 188–200. [CrossRef]
30. Meng, S.; Hsu, Y.C.; Ye, Y.; Chen, V. Submerged membrane distillation for inland desalination applications. *Desalination* **2015**, *361*, 72–80. [CrossRef]
31. Zou, T.; Dong, X.L.; Kang, G.D.; Zhou, M.Q.; Li, M.; Cao, Y.M. Fouling behavior and scaling mitigation strategy of CaSO4 in submerged vacuum membrane distillation. *Desalination* **2018**, *425*, 86–93. [CrossRef]
32. Julian, H.; Lian, B.Y.; Li, H.Y.; Liu, X.F.; Wang, Y.; Leslie, G.; Chen, V. Numerical study of $CaCO_3$ scaling in submerged vacuum membrane distillation and crystallization (VMDC). *J. Membr. Sci.* **2018**, *559*, 87–97. [CrossRef]
33. Julian, H.; Ye, Y.; Li, H.Y.; Chen, V. Scaling mitigation in submerged vacuum membrane distillation and crystallization (VMDC) with periodic air-backwash. *J. Membr. Sci.* **2018**, *547*, 19–33. [CrossRef]
34. Soukanea, S.; Leeb, J.G.; Ghaffourb, N. Direct contact membrane distillation module scale-up calculations: Choosing between convective and conjugate approaches. *Sep. Purif. Technol.* **2019**, *209*, 279–292. [CrossRef]
35. Luo, A.; Lior, N. Study of advancement to higher temperature membrane distillation. *Desalination* **2017**, *419*, 88–100. [CrossRef]
36. Yao, M.W.; Woo, Y.C.; Ren, J.W.; Tijing, L.D.; Choi, J.S.; Kim, S.H.; Shon, H.K. Volatile fatty acids and biogas recovery using thermophilic anaerobic membrane distillation bioreactor for wastewater reclamation. *J. Environ. Manag.* **2019**, *231*, 833–842. [CrossRef] [PubMed]
37. Choi, Y.; Naidu, G.; Lee, S.; Vigneswaran, S. Effect of inorganic and organic compounds on the performance of fractional submerged membrane distillation-crystallizer. *J. Membr. Sci.* **2019**, *582*, 9–19. [CrossRef]
38. Ndinisa, N.V.; Fane, A.G.; Wiley, D.E. Fouling control in a submerged flat sheet membrane system: Part I—Bubbling and hydrodynamic effects. *Sep. Sci. Technol.* **2006**, *41*, 1383–1409. [CrossRef]
39. Kujawa, J.; Kujawski, W. Driving force and activation energy in air-gap membrane distillation process. *Chem. Pap.* **2015**, *69*, 1438–1444. [CrossRef]
40. Singh, D.; Li, L.; Obusckovic, G.; Chau, J.; Sirkar, K.K. Novel cylindrical cross-flow hollow fiber membrane module for direct contact membrane distillation-based desalination. *J. Membr. Sci.* **2018**, *545*, 312–322. [CrossRef]
41. Gustafson, R.D.; Hiibel, S.R.; Childress, A.E. Membrane distillation driven by intermittent and variable-temperature waste heat: System arrangements for water production and heat storage. *Desalination* **2018**, *448*, 49–59. [CrossRef]
42. ZENA Membranes Company. Available online: http://www.zena-membranes.cz/index.php/applications/membrane-distillation-md/water-desalination (accessed on 17 January 2020).
43. Smallwood, J. *Solvent Recovery Handbook*; Edward Arnold: London, UK, 1993; p. 365.
44. Gryta, M.; Tomaszewska, M. Heat transport in the membrane distillation process. *J. Membr. Sci.* **1998**, *144*, 211–222. [CrossRef]
45. McKetta, J.J.; Cunningham, W.A. (Eds.) *Encyclopedia of Chemical Processing and Design*; Marcel Dekker: New York, NY, USA, 1987; Volume 26.
46. Gryta, M. Fouling in direct contact membrane distillation process. *J. Membr. Sci.* **2008**, *325*, 383–394. [CrossRef]
47. Swaminathan, J.; Chung, H.W.; Warsinger, D.M.; Lienhard, J.H. Membrane distillation model based on heat exchanger theory and configuration comparison. *Appl. Energy* **2016**, *184*, 491–505. [CrossRef]
48. M.W. Kellogg Company, and United States. Office of Saline Water. *Saline Water Conversion Engineering Data Book*, 2nd ed.; The M.W. Kellogg Company: Washington, DC, USA, 1971.
49. Gryta, M.; Tomaszewska, M.; Morawski, A.W. A capillary module for membrane distillation process. *Chem. Pap.* **2000**, *54*, 370–374.
50. Zhang, H.Y.; Liu, M.Q.; Sun, D.; Li, B.B.; Li, P.X. Evaluation of commercial PTFE membranes for desalination of brine water through vacuum membrane distillation. *Chem. Eng. Process. Process Intensif.* **2016**, *110*, 52–63. [CrossRef]
51. González, D.; Amigo, J.; Suárez, F. Membrane distillation: Perspectives for sustainable and improved desalination. *Renew. Sustain. Energy Rev.* **2017**, *80*, 238–259. [CrossRef]

52. Khalifa, A.; Ahmad, H.; Antar, M.; Laoui, T.; Khayet, M. Experimental and theoretical investigations on water desalination using direct contact membrane distillation. *Desalination* **2017**, *404*, 22–34. [CrossRef]
53. Park, S.M.; Lee, S. Influence of Hydraulic Pressure on Performance Deterioration of Direct Contact Membrane Distillation (DCMD) Process. *Membranes* **2019**, *9*, 37. [CrossRef]
54. Al-Gharabli, S.; Kujawski, W.; El-Rub, Z.A.; Hamad, E.M.; Kujawa, J. Enhancing membrane performance in removal of hazardous VOCs from water by modified fluorinated PVDF porous material. *J. Membr. Sci.* **2018**, *556*, 214–226. [CrossRef]
55. Zou, T.; Kang, G.D.; Zhou, M.Q.; Li, M.; Cao, Y.M. Submerged vacuum membrane distillation crystallization (S-VMDC) with turbulent intensification for the concentration of NaCl solution. *Sep. Purif. Technol.* **2019**, *211*, 151–161. [CrossRef]
56. Qu, D.; Qiang, Z.M.; Xiao, S.H.; Liu, Q.X.; Lei, Y.Q.; Zhou, T.T. Degradation of Reactive Black 5 in a submerged photocatalytic membrane distillation reactor with microwave electrodeless lamps as light source. *Sep. Purif. Technol.* **2014**, *122*, 54–59. [CrossRef]
57. Ryu, S.; Naidu, G.; Johir, M.A.H.; Choi, Y.; Jeong, S.; Vigneswaran, S. Acid mine drainage treatment by integrated submerged membrane distillation-sorption system. *Chemosphere* **2019**, *218*, 955–965. [CrossRef] [PubMed]
58. Gryta, M. Effectiveness of Water Desalination by Membrane Distillation Process. *Membranes* **2012**, *2*, 415–429. [CrossRef] [PubMed]

Article

Concentrating of Sugar Syrup in Bioethanol Production Using Sweeping Gas Membrane Distillation

Mohammad Mahdi A. Shirazi [1,2] and Ali Kargari [2,*]

1 Membrane Industry Development Institute, Tehran 1587856614, Iran; mmahdiashirazi@gmail.com
2 Membrane Processes Research Laboratory (MPRL), Department of Chemical Engineering, Amirkabir University of Technology (Tehran Polytechnic); Tehran 1591634311, Iran
* Correspondence: kargari@aut.ac.ir

Received: 17 March 2019; Accepted: 25 April 2019; Published: 1 May 2019

Abstract: Membrane distillation (MD) is a relatively new and underdeveloped separation process which can be classified as a green technology. However, in order to investigate its dark points, sensitivity analysis and optimization studies are critical. In this work, a number of MD experiments were performed for concentrating glucose syrup using a sweeping gas membrane distillation (SGMD) process as a critical step in bioethanol production. The experimental design method was the Taguchi orthogonal array (an L_9 orthogonal one) methodology. The experimental results showed the effects of various operating variables, including temperature (45, 55, and 65 °C), flow rate (200, 400, and 600 ml/min) and glucose concentration (10, 30, and 50 g/l) of the feed stream, as well as sweeping gas flow rate (4, 10, and 16 standard cubic feet per hour (SCFH)) on the permeate flux. The main effects of the operating variables were reported. An ANOVA analysis showed that the most and the least influenced variables were feed temperature and feed flow rate, each one with 62.1% and 6.1% contributions, respectively. The glucose rejection was measured at 99% for all experiments. Results indicated that the SGMD process could be considered as a versatile and clean process in the sugar concentration step of the bioethanol production.

Keywords: bioethanol; sweeping gas membrane distillation; SGMD; glucose; permeate flux; optimization

1. Introduction

The increasing global energy demand and its ensuing crisis, as well as the highlighting of major environmental challenges in recent years have led to considerable interest for substituting hydrocarbon fuels with renewable and sustainable energy sources [1–4]. Several alternatives have been explored, including a number of carbon-free sources in an attempt to replace hydrocarbon fuels [5].

Among the different renewable energies, biofuels, and in particular bioethanol as a clean fuel, have gained a great deal of attention [5–7]. Bioethanol can be produced from a wide range of renewable materials such as cellulose, algae, sorghum, and corn biomass. Burning bioethanol, either in place of gasoline or in the form of an ethanol–gasoline, can reduce global warming emissions up to 80%. This can entirely eliminate the release of acid rain [8]. Moreover, bioethanol can be mixed with gasoline for transportation purposes. This technique has been widely used in several countries such as Brazil. However, bioethanol processing and production involves a number of steps (e.g., pretreatment, fermentation, recovery, and refining), and it should be noted that bioethanol can be more significantly beneficial for the environment with improvements to this process, by reducing the required amount of energy. Membrane processes have gained a great deal of attention for their various applications [9,10] including its role in bioethanol processing [11]. This is attributed to their lower energy requirements, lower labor costs, less use of land, and remarkable operational flexibility.

Every bioethanol generation procedure involves saccharification and fermentation processes [5]. However, the concentration of the fermentable sugar in the prehydrolyzates is an important issue. In other words, low sugar concentration can lead to lower bioethanol production, which translates to higher costs and energy consumption in subsequent steps in the process [8]. As a result, the prehydrolyzate should be concentrated (to increase the sugar content) for enhancing the effectiveness of the fermentation step for bioethanol production. Various membrane processes have been used for concentrating sugar syrups such as nanofiltration (NF), reverse osmosis (RO) [12], and membrane distillation [13].

The membrane distillation (MD) process is a relatively new and underdeveloped separation method [10]. This non-isothermal membrane process involves the transport of vapor molecules through the pores of a hydrophobic microporous membrane. Membrane distillation is driven by the vapor pressure difference provided by a temperature difference between the sides of the applied membrane [14]. The hydrophobic characteristic of the applied membranes allows only vapor molecules to pass thorough the pores while holding back the liquid phase. One of the highlighted advantages of the MD process is the relatively considerable permeate fluxes obtained at moderate feed temperatures [15]. In MD, the hydrophobic microporous structure of the membrane plays no role in selectivity for the target component (from a macroscopic point of view) and acts as an interface for the vapor–liquid equilibria (from a microscopic point of view).

In this process, different configurations have been used to impose the driving force and provide the permeate flux. These include direct contact MD (DCMD), air-gap MD (AGMD), vacuum MD (VMD), and sweeping gas membrane distillation (SGMD). Table 1 illustrates a comparison and description of each MD configuration. It is worth noting that, the SGMD seems to be more suitable in processes where permeate is not the target product and can be vented such as concentrating of aqueous solutions containing a non-volatile solute. The application of various MD configurations for different purposes including desalination and bioethanol processing [10,11], the application of atomic force microscopy (AFM) for characterizing the MD membranes [16], and a comprehensive study on polytetrafluoroethylene (PTFE) membranes for MD desalination [17] have been studied comprehensively.

In this work, the sensitivity analysis of the SGMD process for a new application in bioethanol processing, i.e., concentrating the glucose syrup, was investigated. To optimize the experiments, it was obviously necessary to identify which variables were more influential on the target parameter (i.e., the permeate flux in this work). Thus, the Taguchi experimental design (Qualitec-4) was used. Taguchi's approach first defines a set of orthogonal arrays and second devises a standard method for analysis of results. Two important issues should well be pleased by the Taguchi method including the number of trials and the conditions for each one. One of the most important advantages of this method is the significant reduction in time and number of experiments required for obtaining the optimum operating conditions. Moreover, it determines which variable has more influence, and which has less. Therefore, the optimum level for each factor can be determined. Afterward, the confirmation of the predicted value (permeate flux in this work) should be performed.

Table 1. Introduction and explanation of different configurations of membrane distillation (MD) process [14–17].

Configuration	General Scheme	Specification	Description
DCMD	Fo, Po, Fi, Pi	• Both sides of the membrane are in direct contact with the hot and cooling streams	• High-permeate flux • Low-energy efficiency • The simplest operation • The most used configuration • Highest conduction lost

Table 1. *Cont.*

Configuration	General Scheme	Specification	Description
AGMD	Fo Co Fi Po Ci	• A stagnant air-gap in the distillate side is interposed between the membrane and a highly heat-conductive condensing plate	• Highest energy efficiency • Low distillate flux • The air-gap distance can be 2–10 mm
SGMD	Fo Gas vapor Po Fi Gas(i)	• Cold inert gas or air stream is used as carrier for stripping the molecules	• Mostly practical for concentrating purposes • Condensation happens outside the module
VMD	Fo vapor vacuum Po Fi	• Permeate channel is under vacuum	• Mostly practical for volatiles removal • Vapor molecules are condensed outside the module

2. Materials and Methods

2.1. Materials

A commercial flat sheet hydrophobic membrane made of polytetrafluoroethylene (PTFE) (Millipore, USA) with ~170 cm^2 effective area was used. Table 2 presents the specifications of the applied membrane. Pure water (double distillated water) and analytical reagent grade glucose (purity >99.4%, BASF, Germany) were used for preparation of feed samples with the desired concentrations (10, 30, and 50 g/l). Dried air (after filtration) was selected as the sweeping gas.

Table 2. Characteristics of the membrane applied in this study.

Type	PTFE
Pore size (micron)	0.22
Porosity (%)	70
Thickness (micron)	175
Contact angle (°)	132.5 †
Average roughness (nm)	71 †

† Measured value in this work. PTFE: Polytetrafluoroethylene

2.2. SGMD Experimental Apparatus

The applied membrane was placed in a cross-flow plate and frame module with 130 × 130 mm^2 dimensions. The feed and permeate channels depth were 2 mm. A heating bath equipped with a PID controller (Autonics, Korea) and Pt-100 temperature sensors were used for the feed temperature control. Four sensors were located as close as possible to the inlet and outlet sections of the module and another one was located inside the feed tank. A diaphragm pump (So~Pure, Korea) was used to re-circulate the hot feed in a closed loop of "the feed tank-MD module-feed tank". A compressor (an oil-free GAST compressor,

USA, to ensure an oil-free air stream) supplied the sweeping gas (SG) flow. A cooling system was used to condense the vapor molecules. Figure 1 shows the general scheme of the used MD system.

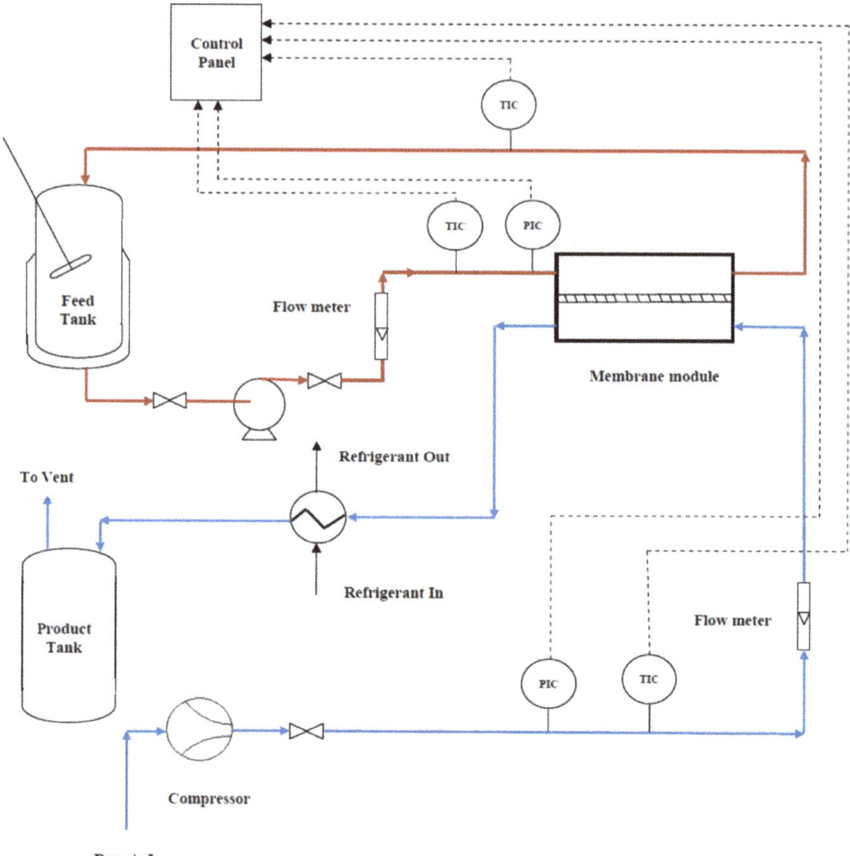

Figure 1. The general scheme of the experimental sweeping gas membrane distillation (SGMD) setup. The system includes the hot side (with red lines): a jacket tank equipped with an over-head mixer and a Pt-100 thermal sensor, a diaphragm pump for recirculation of the feed, flowmeter, the SGMD module equipped with a polytetrafluoroethylene (PTFE) (0.22 µm) membrane and four Pt-100 thermal sensors for entrance and exit points; and the cold side (with blue lines): an oil-free compressor for proving the SG, a flow-meter for adjusting the SG flow, feed tank, and refrigerator system for condensing the permeate vapor.

2.3. Experimental Procedure and Analysis

In each experiment, samples were taken from the feed, permeate, and concentrate every 30 minutes, and analyzed for glucose content by using the glucose oxidase colorimetric method, which has been described in a previous work [18]. The performance of the SGMD process was evaluated based on two major parameters: the permeate flux and the selectivity. Flux was defined as the mass or volume (L) collected in the permeate channel per the membrane's effective area (m^2) and the time (h) of the experiment. Permeate flux and selectivity were calculated using the following equations:

$$F_D = \frac{V}{A \cdot t} \quad (1)$$

$$S = \frac{y_e(1 - x_e)}{x_e(1 - y_e)} \tag{2}$$

where V is the condensed water in the permeate channel, A is membrane's effective area, and t is the time interval, respectively. The y_e and x_e symbols refer to the mass fractions of glucose in the permeate and the feed streams, respectively.

Scanning electron microscopy (SEM) (VEGA, TESCAN, Czech Republic), atomic force microscopy (AFM) (DUALSCOP 95-200E, DEM, Denmark), and contact angle test (KRUSS G-10, Germany) were used for morphology observations of the applied membrane, based on the procedures described in the previous work [16]. Figure 2 shows the SEM and AFM images of the used membrane in this work.

SEM

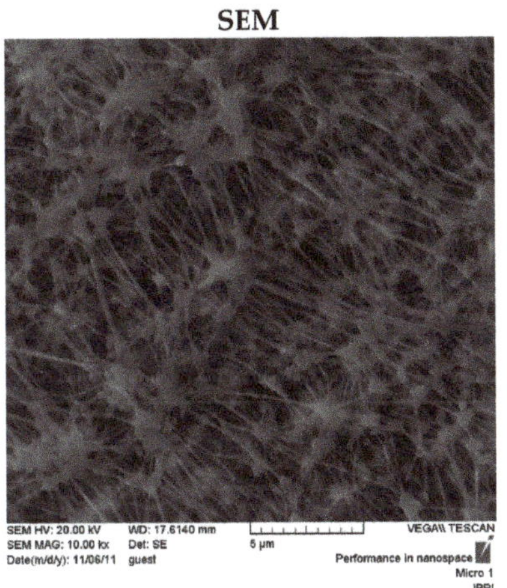

(a) Nominal pore size: 0.22 µm

AFM

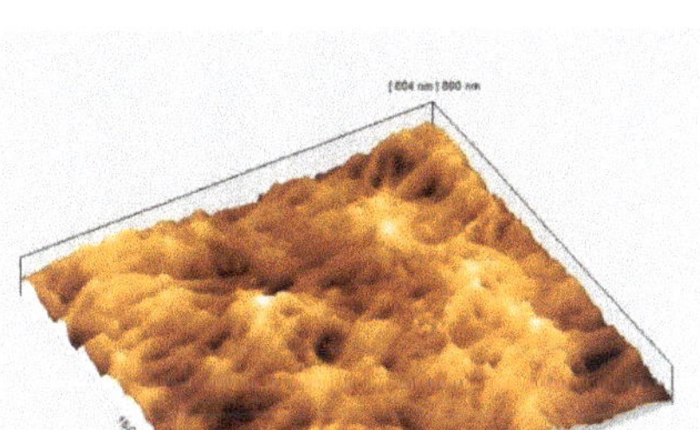

(b) Surface roughness: 71 nm, Surface contact angle: 132.5°

Figure 2. The scanning electron microscopy (SEM) (with 5-µm scale-bar) and atomic force microscopy (AFM) (with 15 µm × 15 µm scanning size) images of the PTFE membrane (with 0.22-µm pore size) used in this work.

3. Results

The steady-state condition of the system was achieved using both distilled water and glucose syrup for about 1 h. The permeate fluxes were reported after this time. An L_9 orthogonal array (four variables in three levels) was offered by Taguchi design methodology. Table 3 represents the experimental variables and their levels. Each row represents a specific experiment. Figure 3a–d shows the main effect of the operating variables on the permeate flux.

Table 3. Operating variables and their levels based on the Taguchi L_9 orthogonal array, as well as the corresponding permeate flux for each test.

Test No.	T_h (°C)	Q_h (ml/min)	C_i (g/l)	Q_a (SCFH)	S (%)	*Flux* (L·m^{-2}·h^{-1})
1	45	200	10	4	25.25	2.98
2	45	400	30	10	31.03	3.54
3	45	600	50	16	24.50	2.58
4	55	200	30	16	55.04	7.36
5	55	400	50	4	29.15	3.22
6	55	600	10	10	41.40	5.19
7	65	200	50	10	68.50	7.16
8	65	400	10	16	82.95	10.36
9	65	600	30	4	42.59	5.72

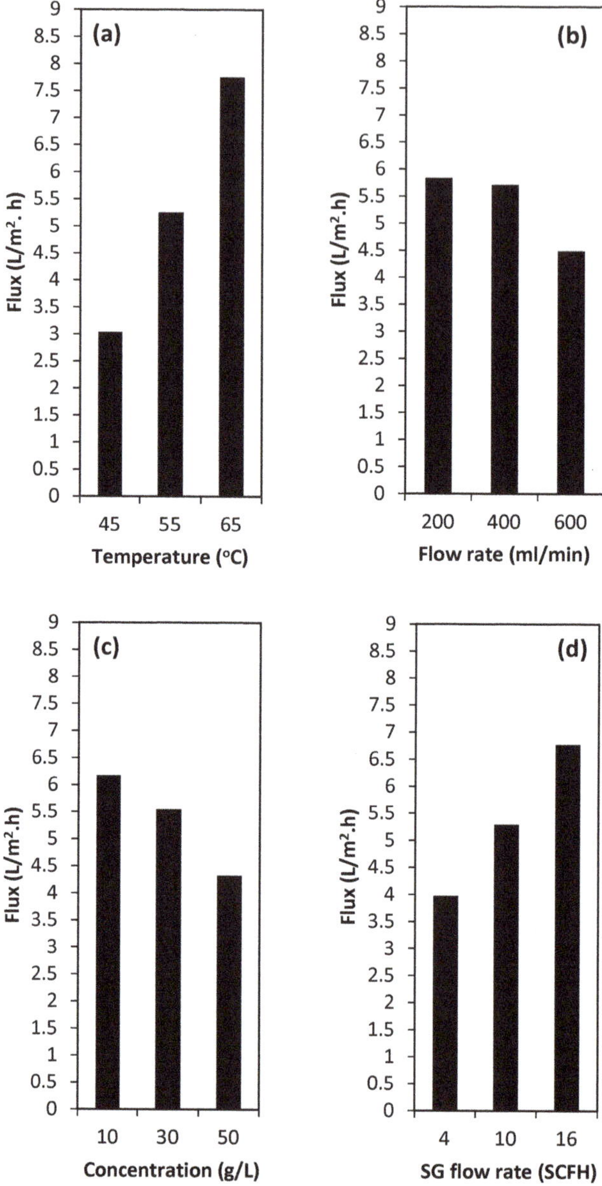

Figure 3. The main effects of feed temperature (**a**), feed flow rate (**b**), feed concentration (**c**), and sweeping gas flow rate (**d**) on the permeate flux based on the operating conditions from the Table 3.

3.1. Main Effects of the Operating Parameters

In this work, the effect of investigated operating parameter of the SGMD process on the permeate flux at different levels (see Table 3) were plotted, separately. This is due to the experimental design which was orthogonal. Figure 3 shows the response value for each level.

Feed temperature (°C) was investigated as the first operating variable. This was due to the nature of the SGMD process which is a non-isothermal separation process. In the SGMD process, the driving force is a function of the temperature difference (ΔT) between the two sides of the membrane pores [19]. As can be observed in Figure 3a, increasing the feed temperature increases the permeate flux, considerably. This can be attributed to the higher vapor pressure in the higher feed temperature. Due to the exponential behavior of the temperature versus the vapor pressure, increasing the feed temperature from 55 to 65 °C proved to have a greater effect than raising it from 45 to 55 °C (see Table 4). It is worth noting that when further increasing the vapor pressure in higher operating temperatures, both the temperature and concentration polarizations increased [20]. Moreover, higher feed temperature can lead to higher heat conduction through the membrane; however, this can be highlighted even more in the DCMD process. Hence, higher temperatures cannot necessarily lead to higher permeate fluxes. Moreover, the results presented in Table 4 indicate that there are some interactions among investigated variables. Therefore, the response of each parameter versus the others should be constructed.

Table 4. Responses for the Taguchi analysis of the permeate fluxes; T_h (feed temperature), Q_h (feed flow rate), C (feed concentration), and Q_a (sweeping gas flow rate).

Responses	Process Variables			
	T_h	Q_h	C_i	Q_a
L_1	3.033	5.833	6.173	3.937
L_2	5.233	5.706	5.539	5.293
L_3	7.746	4.493	4.320	6.766
L_2–L_1	2.22	− 0.127	− 0.635	1.32
L_1–L_2	−2.221	0.126	0.634	−1.321
L_1–L_3	−4.714	1.339	1.825	−2.794
L_2–L_3	−2.494	1.213	1.218	−1.473
L_3–L_1	4.713	−1.340	−1.853	2.793
L_3–L_2	2.493	−1.214	−1.219	1.472

T_h: feed temperature; Q_h: feed flow rate; C: feed concentration; Q_a: sweeping gas flow rate.

Furthermore, the Millard reaction, which is one the major draw-backs in sugar processing [21], is more probable at higher temperatures; hence, the 65 °C was the maximum investigated value for the operating temperature. Moreover, it should be noted that the energy consumption needed to increase the temperature at lower feed temperatures (from 45 to 55 °C in this work) is higher [22,23]; therefore, based on the available energy resource, 65 °C was suggested as the most sufficient operating temperature for concentrating the glucose syrup.

As stated before, the SGMD process is a vapor pressure driven separation. The temperature difference imposes the driving force between the two sides of the membrane, including the feed channel (where the hot feed re-circulates in direct contact with the membrane active layer) and the permeate channel (where vaporization of liquid molecules were carried out when they crossed the membrane pores). Therefore, if the feed flow rate increases too much, the heat transfer between hot feed and cold air will also increase. This means the feed temperature decreases and the temperature of the sweeping air increases [24]. Consequently, this can cause a reduction in the permeate flux [25], and this is also experimentally confirmed in this work (see Figure 3b). As most SGMD experiments carried out using hydrophobic porous membranes were specifically manufactured for microfiltration (MF) applications, designing and developing specific membranes for SGMD applications can solve this draw-back. Although less membrane fouling is expected by using higher feed flow rates, it may decrease the permeate flux, as is shown in Figure 3. This also can be explained by the fact that higher feed flow rates under the constant feed channel depth needs higher inlet pressure. This higher inlet pressure can be translated to the fact of higher pore wetting risk, which consequently can lead to permeate flux decline.

In general, high concentration of solute in the feed stream has an almost negative effect on membrane separations due to the increase in the concentration polarization [26] and vapor pressure reduction (due to the reduction of water activity in aqueous solutions [27]) in MD process. Like other membrane separations, the SGMD process is also sensitive to the concentration polarization, as discussed in the literature [19]. Although this sensitivity is less in the case of the SGMD process; however, the effect of feed concentration on the permeate flux should be studied. As observed in the results of the present work, with increasing the feed concentration, the permeate flux decreases, slightly. This can be attributed to the vapor pressure decline with increasing the glucose concentration in the feed stream. Moreover, higher feed concentration can highlight the effect of the concentration polarization layer (see Figure 3c). It should be noted that one of the most important advantages of the SGMD process in comparison with other membrane separations, which use pressure difference for concentrating sugar syrups, such as the ultrafiltration (UF) and reverse osmosis (RO) process [26], is that the MD process is not sensitive to osmotic pressure limitations. Even though the flux reduction at higher feed concentrations occurs, the MD process can be used for dewatering of the highly concentrated sugar syrups. Figure 3c presents the effect of feed concentration of the target parameter, i.e., the permeate flux.

Figure 3d shows the main effect of the sweeping gas (SG) flow rate on the permeate flux. As can be seen, the SG flow rate can affect the permeate flux, significantly. The higher the sweeping gas flow rates, the higher the fluxes achieved (see Figure 3d). This could be explained as follow. Higher SG flow rate can lead to the vapor pressure reductions in the cold stream (permeate side), which can impose higher mass transfer (larger driving force). Moreover, increasing the SG flow rates can significantly decrease the temperature polarization effect in the cold stream of the MD module. As in the SGMD process, there is no pore wetting risk from the permeate side, the SG flow rate can be as high as possible. However, based on the obtained results it can be recommended that the gas inlet pressure should be lower than that of the hot stream inlet pressure.

3.2. Interactions of Variables

The experimental results indicated that there are some interactions between the parameters. To identify the interaction between the parameters, the response of each parameter against the others should be plotted [27]. The results of interactions are shown in Figure 4. As could be observed, the regions in which there are interactions are illustrated. In each graph in which the lines with different colors (three level for each parameter) cross each other there is an interaction between the parameters.

These interactions are the result of the polarization phenomena in the system. Moreover, the module design can significantly affect the interaction among operating variables. In this work, a plate and frame module with 130 × 130 mm^2 dimensions was used for the SGMD experiments. In some cases, the investigated geometry can make these interactions severer, especially when the DCMD configuration is used because the both sides are in direct contact with the process liquids (temperature polarization is probable in both feed and permeate sides). However, for the SGMD configuration, a lower temperature polarization effect is expected on the permeate side, while the negative effect on the feed side can still exist.

(T_h: feed temperature, Q_h: feed flow rate, C: feed concentration, and Q_a: sweeping gas flow rate).

The results of the interaction study (see Figure 4) show that the higher interaction level was observed for Q_h (feed flow rate) and C (feed concentration) with the severity index (SI) of 71.97%. This can be attributed to the effects of concentration polarization, temperature polarization, and heat loss through the membrane body (the thermal conduction). It can be concluded that the Q_h and C had almost the same effect on the system response (permeate flux). Based on the obtained results, low feed flow rate, and high feed concentration, both can intensify the effect of concentration and temperature polarizations, as well as conductive heat loss through the membrane's body. The interaction between T_h (feed temperature) and Q_h with the severity index of 30.2% is also the result of the temperature polarization. This is true for the other interactions.

On the other hand, the minimum interaction imposes between the feed temperature and the SG flow rate (SI = 8.99%). The feed temperature imposes its effect in the feed channel while the SG flow rate affects the permeate side and these regions are separated by the applied membrane. However, both of these parameters are effective on the permeate flux, separately.

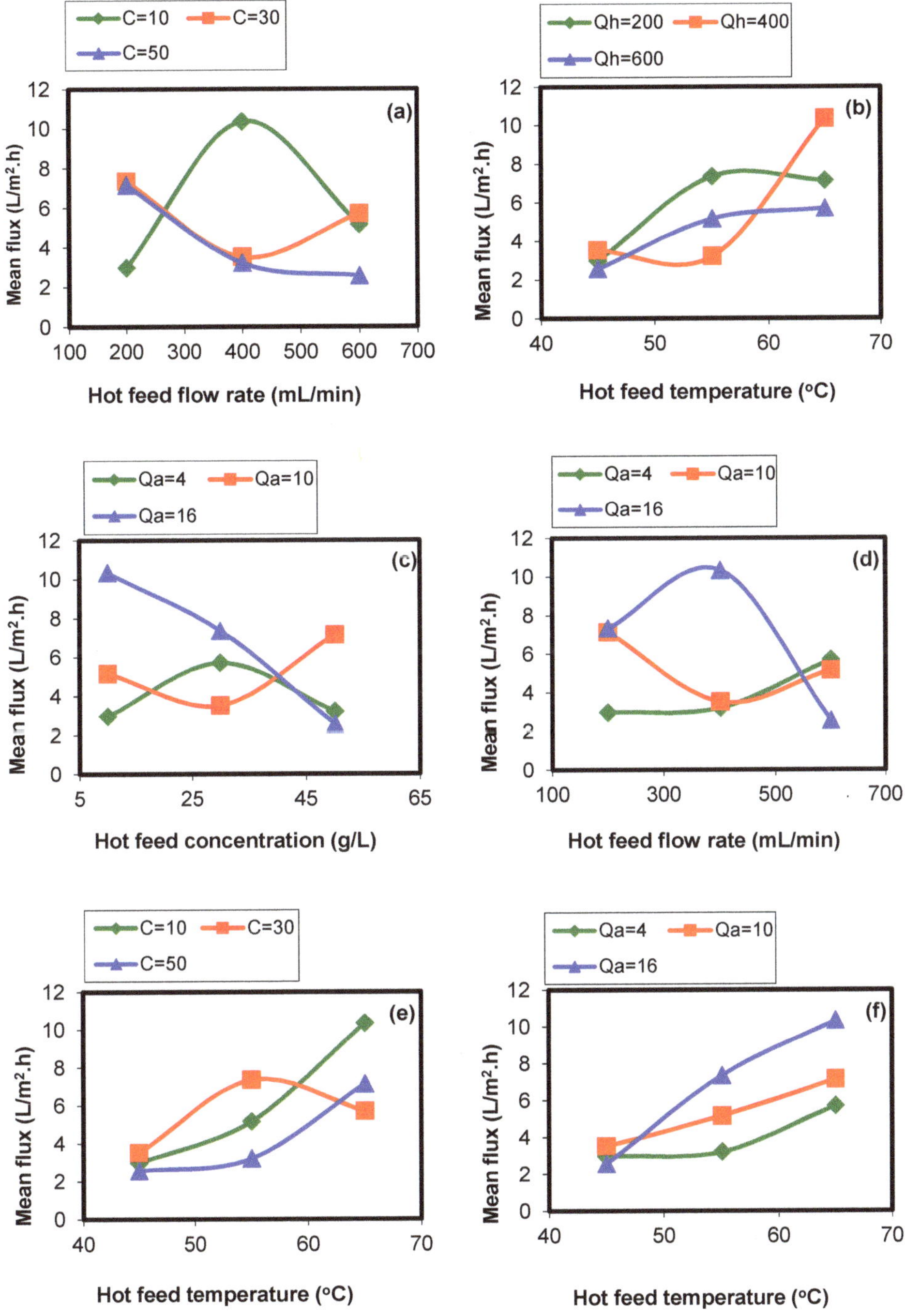

Figure 4. *Cont.*

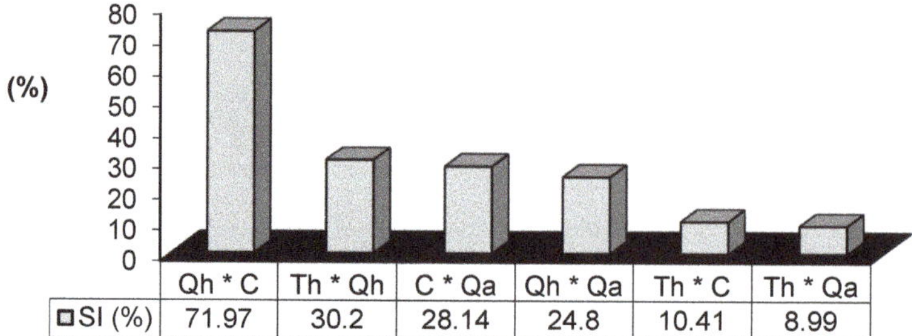

Figure 4. Interactions between the operating variables ((**a**): Q_h versus C; (**b**): T_h versus Q_h; (**c**): C versus Q_a; (**d**): Q_h versus Q_a; (**e**): T_h versus C; and (**f**): T_h versus Q_a) and their related severity index (SI%) based on the operating conditions from Table 3.

3.3. Analysis of Variance (ANOVA)

Using the analysis of variance (ANOVA), the effect of operating parameters on the permeate flux can be investigated, significantly. This is followed up by separating the total variability of each level. This is measured by the sum of the squared (S) deviations from the total mean of the responses, into contribution by each SGMD process parameter and the error [28]. In addition, the importance of SGMD process parameters can be evaluated by the percentage contribution by each of the process parameters in the total sum of the squared deviations.

According to the ANOVA result for the operating parameters of the SGMD process the most significant operating variable(s) affecting the performance characteristic (i.e., the permeate flux as the target parameter in this work) can be investigated. The ANOVA results, which have been shown in Table 5 indicate that the feed temperature and the feed flow rate are the most and the least significant operating variables due to their higher and lower percentages of contribution (62.14% and 6.11%, respectively). As could be observed, the degree of freedom (DOF) for all variables is 2 and F-ratio is zero, based on the obtained results and experimental design in this study.

Moreover, based on the pool-factor analysis, all studied operating variables were effective. The error for these experiments was very low which indicates the accuracy of the performed tests and obtained experimental results.

Table 5. The results of analysis of variance (ANOVA).

Factor	DOF (f)	Sum of Squares	Variance	Contribution Percent P (%)
T_h	2	33.36	16.68	62.14
Q_h	2	3.28	1.641	6.11
C	2	5.32	2.66	9.91
Q_a	2	11.71	5.85	21.82
Other/Error	0	-	-	0.003
Total	8	53.68	-	99.99%

T_h: Feed temperature; Q_h: Feed flow rate; C: Feed concentration; and Q_a: Sweeping gas flow rate.

3.4. Taguchi Model Validation

The optimum operating conditions can be predicted using the Taguchi method after analyzing the proposed experimental results. Moreover, it reports the expected target parameter, i.e., permeate flux, at the proposed optimum operating conditions. Table 6 shows the results of this work. According to Taguchi method, the best combination of the SGMD process parameters is $A_3B_1C_1D_3$. Using this combination of the operating parameters, the highest permeate flux will be attainable. The $A_3B_1C_1D_3$ refers to feed temperature of 65 °C, feed flow rate of 200 ml/min, sugar concentration of 10 g/l, and SG flow rate of 16 SCFH, respectively. The predicted permeate flux under the proposed conditions was measured at 10.48 L/m^2·h. In order to assure the validity of the prediction of the Taguchi model, three individual experiments were carried out using the $A_3B_1C_1D_3$ experimental conditions in different times. Figure 5 shows the results of the Taguchi analysis prediction and the conducted experiments. Although a good agreement between the Taguchi prediction and the conducted experimental results can be observed (see Figure 5), the obtained experimental permeate fluxes were slightly lower than that of the Taguchi prediction. Under the real operating conditions and at different test times there were some noises from the environment, such as humidity and temperature variation which can affect the permeate flux. These noses were investigated by the Taguchi method. That is why the predicted permeate flux was slightly higher than that of the test results.

Table 6. Optimized operating conditions and predicted distillate flux based on the Taguchi method.

Factor	Level	Value
T_h (A)	3	65 °C
Q_h (B)	1	200 mL/min
C (C)	1	10 g/L
Q_a (D)	3	16 SCFH
Expected result for permeate flux at optimum conditions		10.484 (L/m^2·h)

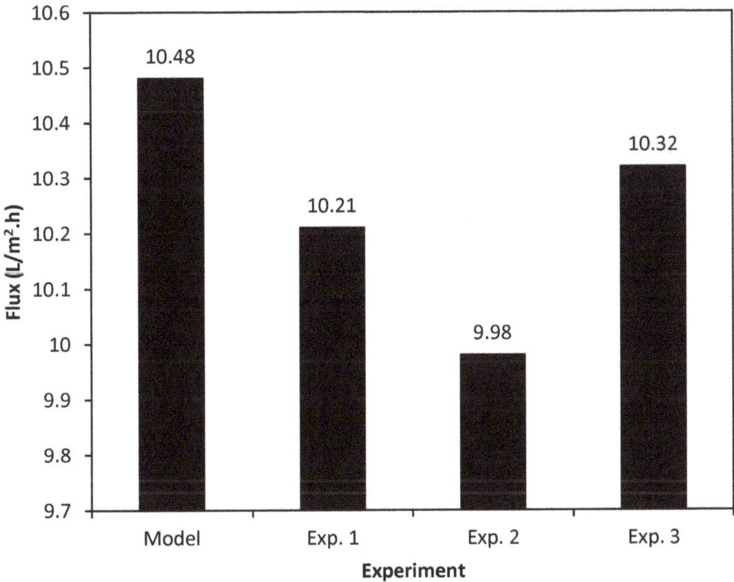

Figure 5. The average permeate flux based on the results of the Taguchi model and experiments after 8 h under the optimum operating conditions (T_h: 65 °C, Q_h: 200 mL/min, C: 10 g/L, and Q_a: 16 SCFH).

Furthermore, Table 7 shows the final concentrations and separation percentages for each experiment after 3 h of operations.

Table 7. Experimental results of this work and literature data for concentrating of sugar syrup using the MD processes.

Year	Feed	Configuration	Conditions	Flux ($L{\cdot}m^{-2}{\cdot}h^{-1}$)	Ref.
1999	Sucrose syrup	AGMD	• T_h: 42.5 °C • Feed flow rate: 43.3 L/h • C_i: 90–300 g/L • Membrane: PVDF (0.2 and 0.45 μm) and PTFE (0.2 and 0.45 μm) • Module: Plate and frame	8.3	[29]
2012	Lignocellulosic hydrolyzates (Glucose)	VMD	• T_h: 65 °C • Q_h: 1 m/s • P: 5 kPa • Membrane: PVDF (0.18 μm) • Module: Hollow-fiber	8.46	[30]
2019	Glucose syrup	SGMD	• T_h: 65 °C • Q_h: 400 mL/min • SG flow rate: 16 SCFH • Membrane: PTFE (0.22 μm) • Module: Plate and frame	10.36	Present work

3.5. SGMD Performance Evaluation

The results of experiments including the separation factor (%S) and the obtained flux ($L{\cdot}m^{-2}{\cdot}h^{-1}$) for each test are summarized in Table 3. As can be observed, the lowest and the highest separation factors were achieved for the lowest glucose concentration in the feed (10 g/L), i.e., 25.25% (test 1) and 82.95% (test 8), respectively. The results confirm that the feed temperature was the most effective operating parameter. This is in good agreement with the literature [29–32]. The highest investigated operating temperature in this study was 65 °C. This temperature was not only in the safe range for processing the sugar syrup (due to thermal sensitivity of the glucose), but was also achieved by using the waste heat in the industrial sectors or by solar heating. The ability of coupling waste heat or solar energy is one of the most important advantages of all MD configurations. Moreover, the 99% solute rejection was achieved for experimental tests in this study.

Table 7 compares the results of this work and the results of the literature. It should be noted that the some of these studies have used the sucrose syrup as the feed sample. Among the MD configurations, the DCMD is the most investigated one for different applications, specifically desalination [32]. However, due to the presence of process liquids in both the feed and permeate channels, the pore wetting risk is considerable [33]. The same concern is attributed to the VMD process, while this configuration is more practical for removing volatile components. On the other hand, the obtained permeate flux using the AGMD process was not high enough for the concentration of the sugar syrup (see Table 7). However, in the SGMD process, only one side of the applied membrane was in direct contact with the process liquid, i.e., in the feed channel. This can considerably lower the pore wetting risk, and consequently make the SGMD process a promising alternative for separations in which the permeance is not the target product. Comparing the results of published data from the literature for concentrating different sugar syrups can also confirm this conclusion. As can be observed in Table 7, the permeate flux of the SGMD configuration was more affordable compared with the other MD configurations.

4. Conclusions

In this work, the Taguchi experimental design and optimization were investigated for the sensitivity analysis of concentrating the glucose syrups using the SGMD process. Feed temperature and sweeping gas flow rate showed the main effects on the permeate flux. Based on the available energy source, feed

temperature at the range of 55 °C to 65 °C is suggested for sugar processing. Results indicated that increasing the feed concentration can lead to a reduction in the permeate flux. Based on the pool-factor analysis, all selected variables were effective. The Taguchi method predicted that the best operating conditions could be achieved using the third, the first, and again the third levels for temperature, feed flow rate and concentration, and sweeping gas flow rate, respectively. Overall, it can be concluded that the SGMD process with a high level of solute rejection (99% in this work for all conducted tests) can be effectively used for the sugar syrup concentration step in bioethanol processing.

Author Contributions: Conceptualization, M.M.A.S. and A.K.; Methodology, M.M.A.S.; Software, M.M.A.S.; Validation, A.K.; Formal Analysis, M.M.A.S.; Writing-Original Draft Preparation, M.M.A.S.; Supervision, A.K.

Funding: This research received no external funding.

Conflicts of Interest: The authors declare no conflict of interest.

Nomenclature

MD	Membrane distillation
SGMD	Sweeping gas membrane distillation
T_h	Feed temperature (°C)
Q_h	Feed flowrate (mL/min)
C_i	Feed concentration (g/L)
Q_a	Sweeping gas flowrate (SCFH)
PTFE	Polytetrafluoroethylene
F_D	Permeate flux ($L{\cdot}m^{-2}{\cdot}h^{-1}$)
S	Selectivity (%)
SG	Sweeping gas
SI	Severity index (%)

References

1. Mekhilef, S.; Saidur, R.; Safari, A. A review on solar energy use in industries. *Renew. Sustain. Energy Rev.* **2011**, *15*, 1777–1790. [CrossRef]
2. Herring, H. Energy efficiency-a critical review. *Energy* **2006**, *31*, 10–20. [CrossRef]
3. Shirazi, M.M.A.; Kargari, A.; Tabatabaei, M.; Mostafaei, B.; Akia, M.; Barkhi, M.; Shirazi, M.J.A. Acceleration of biodiesel-glycerol decantation through NaCl-assisted gravitational settling: A strategy to economize biodiesel production. *Bioresour. Technol.* **2013**, *134*, 401–406. [CrossRef] [PubMed]
4. Hong, S.; Bradshaw, C.J.A.; Brook, B.W. Evaluating options for the future energy mix of Japan after the Fukushima nuclear crisis. *Energy Policy* **2013**, *56*, 418–424. [CrossRef]
5. Taherzadeh, M.J.; Karimi, K. Pretreatment of lignocellulosic wastes to improve ethanol and biogas production: A review. *Int. J. Mol. Sci.* **2008**, *9*, 1621–1651. [CrossRef] [PubMed]
6. Pereira, S.R.; Portugal-Nunes, D.J.; Evtuguin, D.V.; Serafim, L.S.; Xavier, M.R.B. Advances in ethanol production from hardwood spent sulphite liquor. *Process Biochem.* **2013**, *48*, 272–282. [CrossRef]
7. Kocar, G.; Civas, N. An overview of biofuels from energy crops: Current status and future prospects. *Renew. Sustain. Energy Rev.* **2013**, *28*, 900–916. [CrossRef]
8. Wei, P.; Cheng, L.H.; Zhang, L.; Xu, X.H.; Chen, H.; Gao, C. A review of membrane technology for bioethanol production. *Renew, Sustain. Energy Rev.* **2014**, *30*, 388–400. [CrossRef]
9. Takht Ravanchi, M.; Kaghazchi, T.; Kargari, A. Application of membrane separation processes in petrochemical industry: a review. *Desalin.* **2009**, *235*, 199–244. [CrossRef]
10. Shirazi, M.M.A.; Kargari, A.; Bastani, D.; Fatehi, L. Production of drinking water from seawater using membrane distillation (MD) alternative: direct contact MD and sweeping gas MD approaches. *Desal. Water Treat.* **2014**, *52*, 2372–2381. [CrossRef]
11. Shirazi, M.M.A.; Kargari, A.; Tabatabaei, M. Sweeping gas membrane distillation (SGMD) as an alternative for integration of bioethanol processing: study on a commercial membrane and operating parameters. *Chem. Eng. Commun.* **2014**, *202*, 457–466. [CrossRef]

12. Madaeni, S.S.; Zereshki, S. Energy consumption for sugar manufacturing. Part I: Evaporation versus reverse osmosis. *Energy Conver. Manag.* **2010**, *51*, 1270–1276. [CrossRef]
13. Cojocaru, C.; Khayet, M. Sweeping gas membrane distillation of sucrose aqueous solutions: Response surface modeling and optimization. *Sep. Purif. Technol.* **2011**, *81*, 12–24. [CrossRef]
14. Khayet, M. Membranes and theoretical modeling of membrane distillation: A review. *Adv. Colloid Interface Sci.* **2011**, *164*, 56–88. [CrossRef]
15. Shirazi, M.M.A.; Kargari, A.; Shirazi, M.J.A. Direct contact membrane distillation for seawater desalination. *Desal. Water Treat.* **2012**, *49*, 368–375. [CrossRef]
16. Shirazi, M.M.A.; Kargari, A.; Tabatabaei, M.; Ismail, A.F.; Matsuura, T. Assessment of atomic force microscopy for characterization of PTFE membranes for membrane distillation (MD) process. *Desal. Water Treat.* **2014**, *54*, 295–304. [CrossRef]
17. Shirazi, M.M.A.; Kargari, A.; Tabatabaei, M. Evaluation of commercial PTFE membranes in desalination by direct contact membrane distillation. *Chem. Eng. Process.* **2014**, *76*, 16–25. [CrossRef]
18. Sohrabi, M.; Kargari, A. Modeling and application of UF-stirred cell reactors in hydrolysis of saccharides. *Chem. Eng. Technol.* **1999**, *22*, 955–962. [CrossRef]
19. Alkhudhiri, A.; Darwish, N.; Hilal, N. Membrane distillation: A comprehensive review. *Desalination* **2011**, *287*, 2–18. [CrossRef]
20. Shirazi, M.M.A.; Kargari, A.; Tabatabaei, M.; Ismail, A.F.; Matsuura, T. Concentrating of glycerol from dilute glycerol wastewater using sweeping gas membrane distillation. *Chem. Eng. Process.* **2014**, *78*, 58–66. [CrossRef]
21. Krishnapura, P.R.; Belur, P.D.; Subramanya, S. A critical review on properties and applications of microbial L–asparaginases. *Critical Rev. Microbiol.* **2016**, *42*, 720–737. [CrossRef]
22. El-Dessouky, H.T.; Ettouney, H.M.; Mandani, F. Performance of parallel feed multiple effect evaporation system for seawater desalination. *Appl. Thermal Eng.* **2000**, *20*, 1679–1706. [CrossRef]
23. Kronenberg, G.; Lokiec, F. Low-temperature distillation processes in single- and dual-purpose plants. *Desalination* **2001**, *136*, 189–197. [CrossRef]
24. Rivier, C.A.; Garcia-Payo, M.C.; Marison, I.W.; von Stockar, U. Separation of binary mixtures by thermostatic sweeping gas membrane distillation: I. Theory and simulation. *J. Membr. Sci.* **2002**, *201*, 1–16. [CrossRef]
25. Garcia-Payo, M.C.; Rivier, C.A.; Marison, I.W.; von Stockar, U. Separation of binary mixtures by thermostatic sweeping gas membrane distillation: II. Experimental results with aqueous formic acid solutions. *J. Membr. Sci.* **2002**, *198*, 197–210. [CrossRef]
26. Shirazi, S.; Lin, C.J.; Chen, D. Inorganic fouling of pressure-driven membrane processes- A critical review. *Desalin.* **2010**, *250*, 236–248. [CrossRef]
27. Hitsov, I.; Maere, T.; De Sitter, K.; Dotremont, C.; Nopens, I. Modelling approaches in membrane distillation: A critical review. *Sep. Purif. Technol.* **2015**, *142*, 48–64. [CrossRef]
28. Kargari, A.; Kaghazchi, T.; Sohrabi, M.; Soleimani, M. Batch extraction of gold(III) ions from aqueous solutions using emulsion liquid membrane via facilitated carrier transport. *J. Membr. Sci.* **2004**, *233*, 1–10. [CrossRef]
29. Izquierdo-Gil, M.A.; Garcia-Payo, M.C.; Fernandez-Pineda, C. Air gap membrane distillation of sucrose aqueous solutions. *J. Membr. Sci.* **1999**, *155*, 291–307. [CrossRef]
30. Zhang, L.; Wang, Y.; Cheng, L.H.; Xu, X.; Chen, H. Concentration of lignocellulosic hydrolyzates by solar membrane distillation. *Bioresour. Technol.* **2012**, *123*, 382–385. [CrossRef]
31. Al-Asheh, S.; Banat, F.; Qtaishat, M.; Al-Khateeb, M. Concentration of sucrose solutions via vacuum membrane distillation. *Desalin.* **2006**, *195*, 60–68. [CrossRef]
32. Ullah, R.; Khraisheh, M.; Esteves, R.J.; McLeskey, J., Jr.; AlGhouti, M.; Gad-el-Hak, M.; Vahedi Tafreshi, H. Energy efficiency of direct contact membrane distillation. *Desalin.* **2018**, *433*, 56–67. [CrossRef]
33. Zare, S.; Kargari, A. Membrane properties in membrane distillation. In *Emerging Technologies for Sustainable Desalination Handbook*, 1st ed.; Gude, V.G., Ed.; Elsevier: Amsterdam, The Netherlands, 2018; pp. 107–156.

 *membranes*

Article

Low-Temperature Direct Contact Membrane Distillation for the Treatment of Aqueous Solutions Containing Urea

Alessandra Criscuoli [1,*], Alfredo Capuano [2], Michele Andreucci [3] and Enrico Drioli [1]

1 Institute on Membrane Technology (ITM-\CNR), via P. Bucci 17/C, 87036 Rende (CS), Italy; e.drioli@itm.cnr.it
2 U.O.C. Nefrologia e Trapianto, A.O.U Federico II, 80131 Napoli, Italy; alfredo.capuano@unina.it
3 Renal Unit–Department of Health Sciences of "Magna Graecia" University–Viale Europa, Campus Salvatore Venuta, 88100 Catanzaro, Italy; andreucci@unicz.it
* Correspondence: a.criscuoli@itm.cnr.it; Tel.: +39-0984-492118

Received: 1 July 2020; Accepted: 31 July 2020; Published: 3 August 2020

Abstract: Research activities on the application of direct contact membrane distillation (DCMD) for processing at low temperature (up to 50 °C) solutions containing urea were presented and discussed. Feeds were urine (also in mixture) and human plasma ultrafiltrate. Moreover, as a case study, the performance of membrane modules of different sizes and features was investigated for reaching the productivities needed in the treatment of the human plasma ultrafiltrate. In particular, two modules were equipped with the same type of capillaries, but differed in terms of membrane area, while the third module contained a different type of membranes and presented a membrane area in between those of the two previous modules. The three modules were compared, at a parity of operating temperatures and streams velocity, in terms of transmembrane flux, permeate production and size, underlining the directions to follow for a real implementation of the technique.

Keywords: direct contact membrane distillation; urea; low temperature

1. Introduction

Direct contact membrane distillation (DCMD) is the most investigated configuration for various applications [1], because it is easy to implement and handle. In fact, the water vapor produced at the feed side is directly condensed at the distillate side, without requiring external condensers, as in vacuum membrane distillation (VMD) and Sweep Gas Membrane Distillation (SGMD). If compared with the air gap membrane distillation (AGMD) configuration, the DCMD operates with simpler membrane modules, in which only membranes are packed, and there is no need of condensing surfaces. In DCMD, one side of a microporous hydrophobic membrane is in direct contact with the feed to be distilled, while the other side contacts the distillate stream. The distillation occurs by applying a difference of temperature between the feed (hot stream) and the distillate (cold stream), resulting in a difference of vapor pressures across the membrane that promotes the transport of the water vapor from the feed to the distillate side through the dry membrane micropores (Figure 1).

During the distillation, the feed stream loses heat because of its evaporation, while the distillate stream warms up, due to the water vapor condensation. Therefore, at the exit of the membrane module, the streams are heated and cooled, respectively, to recycle them back to the membrane unit at the desired temperature values. The typical operating temperatures of the feed range between are 60 and 80 °C, in order to obtain a high difference of vapor pressures and then, high transmembrane fluxes, especially during concentration tests. However, for some applications, a lower temperature must be applied. In addition to the agrofood and pharmaceutical fields, lower operating temperatures are needed, for example, when treating biological fluids to avoid the denaturation and degradation of compounds. This is also the case of streams containing urea, for which most of the DCMD tests were carried out in the range of 40–50 °C. The purification of these streams is of interest for the wastewater treatment in space, for processing the wastewater coming from urea synthesis plants, as well as for the treatment of patients affected by chronic renal failure. In DCMD, it is possible to recover purified water, to re-used as distillate, and to produce a stream concentrated in urea, that can be further used in the production of fertilizers or in resin fabrication. The possibility to re-use the wastewater, including urine, in space is of high importance, because it avoids the need of external water supply and of wastewater storage/disposal [2]. Similarly, the recovery of water from the human plasma ultrafiltrate of patients undergoing extracorporeal blood purification techniques avoids the use of external water as dialysate and/or reinfusate fluid, strongly reducing the risk of inflammatory problems linked to the presence of chemical pollutants (even if in traces) [3]. The aim of this work was first to present and discuss the researches made in processing at low temperature (up to 50 °C) solutions containing urea by DCMD, specifically, urine (also in mixture) and human plasma ultrafiltrate. Then, as a case study, the results of experiments made to improve the productivity of the process for the treatment of the human plasma ultrafiltrate were reported. Tests were carried out on three commercial modules of different size (0.1, 0.35 and 0.83 m^2) equipped with capillary polypropylene membranes (same membrane properties for the 0.1 m^2 and 0.83 m^2 modules). The efficiency of the modules was compared at a parity of operating temperatures and streams velocity, and the module with the best performance was identified.

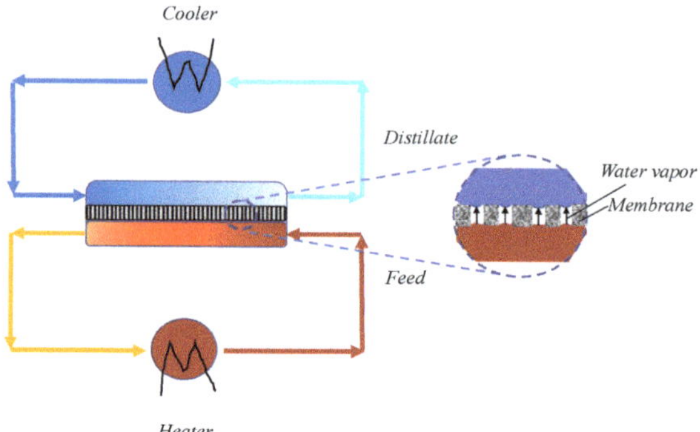

Figure 1. Direct contact membrane distillation (DCMD) process.

2. Research Activities

The concept of DCMD was used for improving the performance of a combined forward osmosis (FO)–osmotic distillation (OD) unit for the treatment of metabolic wastewater in space [2]. Two systems were investigated: one where the MD membrane worked under a difference of temperature only (FO/MD), another where both a difference of temperature and concentration were applied (FO/membrane osmotic distillation (MOD). In the latter case, the permeate stream consisted in an osmotic agent (NaCl solution) rather than distilled water. In both systems, the MD membrane was laid on a semipermeable FO membrane. The synthetic wastewater was prepared including the main sources of wastewater on a spacecraft, like hygiene wastewater, humidity condensate and urine (urea, 5 g/L wastewater). The MD membrane was in flat configuration and made of polypropylene (PP; 0.22 μm of pore size, GE Osmonics, Minnetonka, MN). Tests on FO/MD were carried out on a solution of urea in deionized water and on a triply concentrated synthetic wastewater. The operating temperatures were kept at 25 and 21 °C for the feed and permeate, respectively. A constant flux of 0.8 L/m^2 h was obtained for both feeds and a concentration factor of 9 was registered for the wastewater after about 70 h. Neither urea nor surfactant were found in the permeate, confirming the good rejections of the MD and FO membrane, respectively. The FO/MOD unit was tested at the same operating conditions of the FO/MD, but sending at the permeate side NaCl solutions (60–100 g/L). The flux was constant and around 0.9–1 L/m^{2}h and also in this case a complete rejection of urea was obtained. In 15 days of test, the FO/MD system led to a 4–20 times higher flux than the FO/OD unit, while with the FO/MOD, a 8–25 times higher flux than the FO/OD unit was registered.

Always in the logic of wastewater recycling in space, DCMD tests were carried out by Cartinella et al. [4] on a mixture of humidity condensate and urine by using a commercial capillary module, MD020-CP-2N (Microdyn, Germany), of 0.1 m^2 membrane area. The feed, at 40 °C, was sent inside the capillaries while the distilled water at 20 °C flowed countercurrently at the shell side. At a water recovery factor of 75%, the flux was about 1.5 L/m^{2}h and the urea rejection was higher than 99.9%. Furthermore, the estrone and estradiol rejections were also analyzed and values higher than 99.5% were obtained.

The potential of DCMD coupled to FO was studied by Liu et al. [5] for the treatment of real human urine. The MD membrane was made of polytetrafluoroethylene (PTFE) and had a pore size of 0.45 μm (Jitian Company, Shangai, China). The DCMD unit used distilled water at the permeate side and had the scope of re-concentrating the draw solution (NaCl, 1–2.5 M), coming from the FO, and of producing the final distillate. In 8 h tests, the water transfer rate of the FO/MD system was around 3.39 L/m^2 h at 40 °C and 1M NaCl as draw solution, and increased to 5.08 L/m^2 h at 53 °C and 2.5 M NaCl. The overall rejection was nearly 100% for all contaminants (not only urea), due to the high rejection of MD for the non volatile species and of FO for the volatile ones. The concentrated urine can be used for nutrients recovery.

The integration of FO with DCMD was also investigated by Volpin et al. [6] who optimized the FO and DCMD operating parameters to reduce the nitrogen content in the produced distillate. In particular, for the DCMD unit, the effect of the feed temperature (from 40 to 60 °C) and membrane properties were analyzed, while keeping the cross-flow velocities (8.5 cm/s) and the temperature at the distillate side (20 °C) both constant. The membranes used were from Merk Millipore and had the same nominal pore size (0.22 μm), but differed in terms of material (polyvinylidene fluoride (PVDF) and PTFE), porosity and thickness (see Table 1). At all investigated temperatures, the PTFE membrane led to higher fluxes than the PVDF, due to the higher porosity and contact angle. Specifically, the transmembrane fluxes for the PTFE membrane and a 1.5 M NaCl feed varied from 6 L/m^2 h to 12 L/m^2 h at 40 and 50 °C, respectively, further increasing up to 16 L/m^2 h at 60 °C. At the same operating conditions, the PVDF membrane led to fluxes of 4 L/m^{2}h (40 °C), 7 L/m^2 h (50 °C) and 11 L/m^2 h (60 °C). The PTFE membrane was also able to give water fluxes higher than the NH_3 flux, producing a high-quality distillate.

Microporous hydrophobic composite membranes for water recovery from urine were prepared and tested by Khumalo et al. [7] who modified the PVDF/PTFE membranes with methyl functionalized silica nanoparticles (fMSNs). In particular, three membranes were prepared: M1-PVDF/0.3% fMSNs; M2- PVDF/3% PTFE/0.3% fMSNs; M3- PVDF/6% PTFE/0.3% fMSNs. DCMD tests were made on hydrolyzed human urine that was sent at one side of the membrane at 50 °C while deonised water circulated at the other side at 20 °C. The M3 membrane, with the highest contact angle and a less porous structure, performed better, showing higher rejections towards ammonia–nitrogen (99.1%), TOC (>98%), Na^+ and K^+ (>99%). A water recovery factor of 80% was obtained with fluxes similar to those reported in the literature at the same temperature difference. Nevertheless, the authors pointed out that fouling issues due to the possible deposition of urine compounds on the membrane surface must be taken into account during the treatment.

The application of DCMD to purify the human plasma ultrafiltrate of patients affected by chronic renal failure was investigated by Capuano et al. [3], who used a commercial capillary module, MD020-CP-2N (Microdyn, Germany), of a 0.1 m^2 membrane area. The experiments were made on two synthetic solutions containing urea and NaCl in different amounts (solution A: urea, 2 g/L and NaCl, 9 g/L; solution B: urea, 8 g/L and NaCl, 36 g/L), as well as on real human plasma ultrafiltrate (urea, 1.09 g/L; creatinine, 0.054 g/L; Na^+, 3.27 g/L; K^+, 0.129 g/L; Ca^{2+}, 0.062 g/L). The feed was sent, at different flow rates (6–200 L/h) and temperatures (29–39 °C), inside the capillaries, while osmotic water recirculated counter-currently at the shell side. The highest transmembrane flux was 3 L/m^2 h at 39 °C and 200 L/h of feed flow rate. In all tests, neither urea nor other species in the feed permeated through the membrane and the collected distillates were of high-quality. Furthermore, no fouling issues were registered, also during prolonged tests (lasted 4.5 h and 7 h) carried out on the real plasma ultrafiltrate. However, the permeate produced by the module (0.3 L/h) did not match the distillate needs for a clinical application of the technique, that ranged from 20 to 30 L/h.

Table 1 summarizes the main results related to the research activities discussed. It is evident that only few studies were carried out on the application of low-temperature DCMD, and mainly at the lab-scale (membrane areas between 20 cm^2 and 0.1 m^2). Although the used membranes were mostly commercial, the modules were often lab-made and in flat configuration, with the exception of some tests made on a commercial capillary module [3,4]. The membranes were in PVDF, PTFE or PP, often with a typical pore size of 0.2 μm and a porosity between 70% and 85%, while their thickness varied from 125 to 450 μm. Nevertheless, all studies on the topic demonstrated an excellent rejection of DCMD for urea (up to 100%), confirming its purification and concentration capability. The use of the integrated FO/DCMD system resulted as beneficial for the final permeate quality, due to the FO rejection of volatiles, such as ammonia and nitrogeneous organic species, and surfactants, that preserved DCMD from wetting, combined with the high DCMD rejection of all non-volatiles, included urea.

Table 1. Main results of the research activities on the treatment of solutions containing urea by low-temperature DCMD.

Urea Source	Module Geometry	Membrane Properties	Operating Conditions	Main Results	Refs.
Feed 1: Solution of urea; Feed 2: Triply concentrated synthetic wastewater	Flat	PP; d_p: 0.22 μm; ε: 70%; δ: 150 μm; A_m: 139 cm^2	T_f: 25 °C; T_d: 21 °C; v_f: 0.1 m/s	FO/MD: J: 0.8 L/m^{2}h; R_{urea}: 100%; Conc. factor (70 h): 9 FO/MOD: J: 0.9–1 L/m^{2}h; R_{urea}: 100%	[2]
Synthetic mixture of humidity condensate and urine	Capillary	PP; d_p: 0.2 μm; ε: 70%; δ: 450 μm; A_m: 0.1 m^2	T_f: 40 °C; T_d: 20 °C; Q_f: 90 L/h	Water recovery factor: 75%; J: 1.5 L/m^{2}h; R_{urea}>99.9%	[4]
Real human urine	Flat	PTFE; d_p: 0.45 μm; δ: 180 μm; A_m: 29.5 cm^2	T_f: 40–55 °C; T_d: 25 °C; Q_f: 12 L/h	FO/MD (40 °C; 1 M NaCl): J: 3.39 L/m^{2}h; R_{urea}: 100%	[5]
Synthetic urine	Flat	PVDF; d_p: 0.22 μm; ε: 75%; δ: 125 μm; PTFE; d_p: 0.22 μm; ε: 85%; 150 μm; A_m: 20 cm^2	T_f: 40–60 °C; T_d: 20 °C; v_f: 0.085 m/s	FO/MD (40 °C; 1.5 M NaCl): J_{PVDF}: 4 L/m^{2}h; J_{PTFE}: 6 L/m^{2}h	[6]
Hydrolyzed real human urine	Flat	PVDF/6%PTFE/0.3%fMSNs; Contact angle: 115.5°; A_m: 125 cm^2	T_f: 50 °C; T_d: 20 °C; Q_f: 35 L/h	Water recovery factor: 80%; $R_{ammonia\text{-}nitrogen}$: 99.1%	[7]
Feed 1: synthetic solution of urea and NaCl Feed 2: real human plasma ultrafiltrate	Capillary	PP; d_p: 0.2 μm; ε: 70%; δ: 450 μm; A_m: 0.1 m^2	T_f: 29-39 °C; T_d: 13–20 °C; Q_f: 6–200 L/h	J (39 °C; 200 L/h): 3 L/m^{2}h; R_{urea}: 100%	[3]

d_p: mean membrane pore size; ε: membrane porosity; δ: membrane thickness; A_m: membrane area; T_f: feed temperature; T_d: distillate temperature; v_f: feed velocity; Q_f: feed flow rate; J: transmembrane flux; R: rejection; PP: polypropylene; PTFE: polytetrafluoroethylene; PVDF: polyvinylidenefluoride; FO/MD: forward osmosis/membrane distillation; MOD: membrane osmotic distillation.

3. Case Study: Improvement of the Permeate Production for the Treatment of the Human Plasma Ultrafiltrate by DCMD

Based on the results obtained in our previous work [3] which was presented in Section 2, through research activities, modules with higher membrane areas were investigated in order to increase the permeate production during the treatment of the plasma ultrafiltrate by DCMD. In particular, another module, produced by Microdyn (Wuppertal, Germany), the MD063CP2N, and a tailor made module (PF2000N), based on PF2000N design and containing a non-treated plasmaphan membrane (3M) provided by Gambro Dialysatoren GmbH (Hechingen, Germany), were tested at the same operating conditions. Hereinafter, the three modules will be identified as M1 (MD020CP2N), M2 (MD063CP2N) and M3 (PF2000N).

3.1. Materials and Methods

The M2 module was equipped with the same type of membranes of M1, in polypropylene, but was longer and contained a higher number of fiber. It has to be noticed that the data sheet of the Company reported a number of 200 fibers and a membrane area of 0.75 m^2. However, the effective number of fibers was 220, so the membrane area used for the flux calculation was increased accordingly (0.83 m^2) and also the free flow areas of the lumen and shell side were calculated considering the effective number of fibers. The M3 module contained a higher number of membranes (around 1500) made also of polypropylene, but with different properties and size. Its membrane area was in between those of the two previous modules (0.35 m^2). The main properties of the membranes and modules are reported in Table 2; Table 3, respectively.

Table 2. Main properties of the membranes.

Module	Membrane Material	Pore Size (µm)	Porosity (%)	Thickness (mm)	Inner Diameter (mm)	Outer Diameter (mm)
M1/M2	Polypropylene	0.2	70	0.45	1.8	2.7
M3	Polypropylene	0.3	70	0.15	0.33	0.63

Table 3. Main properties of the modules.

Module	Number of Fibers	Membrane Area (m^2)	T_{max} (°C)	Size (cm × cm)
M1	40	0.1	40	50 × 2.5
M2	220	0.83	40	75 × 6.3
M3	1555	0.35	50	25 × 4.14

Both microdyn modules have 40 °C as the maximum operating temperature, while the M3 module is usually operated at around 37 °C. Therefore, the comparison experiments were carried out at around 37 °C of feed temperature. The permeate temperature was around 22 °C. The maximum operating feed and permeate flow rates that could be used with the M3 module without increasing the relative pressure values were 170 L/h and 250 L/h, respectively. The corresponding velocities were 0.35 m/s for the feed stream and 0.16 m/s for the permeate side. The operating feed and permeate flow rates to be used in the two Microdyn modules were then calculated, considering these velocity values. It has to be noticed that, with the above feed velocities, the three modules worked in laminar regime, with the Reynolds values being equal to 154 and 840 for the M3 and the two microdyn modules, respectively.

3.2. Mass Transfer

In DCMD, the water vapor mass flux is a function of the membrane properties, as well as of the difference of water vapor pressure across the membrane (driving force) and can be described as

$$J = K \times (P_{fm} - P_{dm}) \quad (1)$$

where J (kg/m^2s) is the transmembrane flux, K (kg/m^2 s Pa) is the membrane distillation coefficient, P_{fm} (Pa) is the feed vapor pressure at the membrane surface-hot side, and P_{dm} (Pa) is the distillate vapor pressure at the membrane surface-cold side. The membrane distillation coefficient includes the membrane properties. They are usually assembled in two ratios based on two main mechanisms that can occur separately or in combination, depending on the membrane pore size and operating conditions: Knudsen and molecular diffusion. The type of mechanism occurring in the membrane is usually identified by calculating the Knudsen coefficient as the ratio between the mean free path of molecules and the mean pore size of the membrane. When the Knudsen coefficient is higher than 1, the Knudsen transport dominates, while for Knudsen coefficient values lower than 0.01, the molecular diffusion takes place. For Knudsen coefficient values between 0.01 and 1, both mechanisms can occur (transition region) [8]. The mean free path of water vapor can be calculated by [8]

$$l = \frac{k_B T}{\pi \left(\frac{\sigma_w + \sigma_a}{2}\right)^2 P_{pore} \sqrt{1 + \left(\frac{M_w}{M_a}\right)}} \quad (2)$$

where k_B is the Boltzman constant (J/K), T is the mean temperature in the pores (K), σ_w is the collision diameters of water vapor (m), σ_a is the collision diameters of air (m), P_{pore} is the air pressure in the pores (kPa), M_w is the water molecular weight (g/mol), and M_a is the air molecular weight (g/mol).

3.3. Specific Thermal Energy Consumption (STEC) and Productivity/Size (PS) Ratio

An important aspect in membrane distillation is the thermal energy consumption linked to the heat of the feed stream [9]. In this respect, the STEC is often used, calculated as the energy to supply to the feed recirculating inside the module, divided by the permeate production:

$$STEC = \frac{Q_f \times c_p \times (T_{fin} - T_{fout})}{Q_p} \quad (3)$$

where Q_f is the feed flow rate (kg/h), c_p is the specific heat of the feed (kJ/kg K), T_{fin} is the feed temperature at the module inlet (K), T_{fout} is the feed temperature at the module outlet (K), and Q_p the permeate flow rate (L/h).

Another important parameter is the size of the membrane distillation plant needed for obtaining a certain productivity. In the process intensification strategy, future plants should ensure high productivities and high compactness. In this logic, a metric was defined to compare the plants in terms of productivity and size. In particular, the productivity/size ratio (PS) metric, compares the ratio of the productivity and size of two plants [10]:

$$PS \frac{\text{Productivity}/_{\text{Size}} \mid \text{plant}_1}{\text{Productivity}/_{\text{Size}} \mid \text{plant}_2} \quad (4)$$

When the PS is higher than 1, plant 1 must be preferred, otherwise, plant 2 must be chosen.

3.4. Experimental Set-Up and Procedure

DCMD tests on the different modules were carried out on the same lab set-up.

As already mentioned, the case study was based on the results obtained in our previous work [3], where both artificial solutions containing urea (0.2 wt % and 0.8 wt %) and real ultrafiltrate samples (urea content: 0.1 wt %) were treated by the M1 module. These experiments were carried out in the feed flow rate range of 6–200 L/h, corresponding to Reynolds values of 39 and 1292, respectively. In that study, the urea was completely rejected and no fouling issues were observed, also in prolonged tests on real ultrafiltrate samples. In the case study of the present work, the M2 module was equipped with the same fibers of the M1, so the same behaviour with solutions containing urea is expected. The M3 module was also equipped with fibers made of polypropylene, thus, the interactions between the urea and the membrane material can be considered the same as for the other two modules. Furthermore, the M3 module was operated at the same feed velocities as the M1 and M2 modules and the Reynolds values involved (154 for the M3 and 840 for the Microdyn modules) fell in the range investigated in our previous work. Finally, it has to be noticed that the feed of interest is quite diluted (the human plasma ultrafiltrate contains about 0.1 wt % of urea, that is highly soluble in water). On the basis of the above considerations, the experiments to investigate the possibility of increasing the permeate productivity were carried out using distilled water as both feed and permeate stream.

The two streams were re-circulated in counter-current flow mode to the module (the feed in the lumen and the permeate in the shell), after their heating and cooling, respectively. In the set-up, the module inlet and outlet temperatures were measured for both hot and cold lines and the operating pressures were monitored by two manometers, located at the two entrances of the module. All tests were carried out at atmospheric pressure. The permeate flux was calculated by registering the weight change of the distillate tank, located on an electronic balance, and by dividing the accumulated mass by the membrane area and the operating time. The experiments lasted one hour and no significant changes in the flux values were registered in time. The M1 and M3 modules were used in horizontal position, for an immediate fixing with the existing tubes of the lab set-up, whilst the M2 module, having a bigger size, was mounted in vertical and fed from the bottom with the permeate stream, to ensure that all the shell side was wetted and that no channeling occurred (Figure 2). It has to be mentioned that both the M1 and M3 modules can also be used in vertical position, as M2.

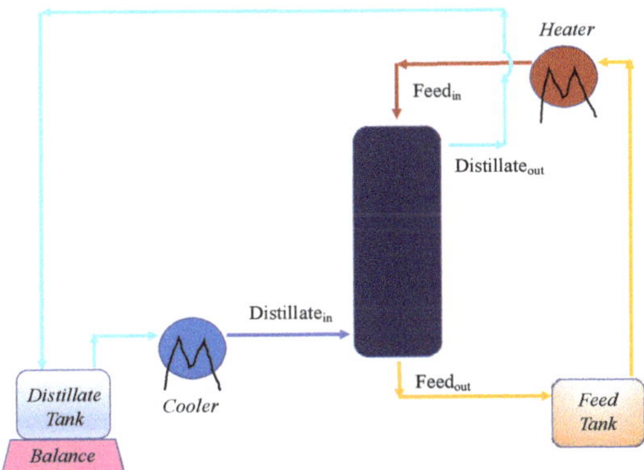

Figure 2. How the streams were fed to the M2 module in the lab set-up.

3.5. Results and Discussion

The transmembrane flux obtained with the three modules operating at the same conditions is shown in Figure 3. A comparison of the modules in terms of productivity (Qp) is reported in Figure 4.

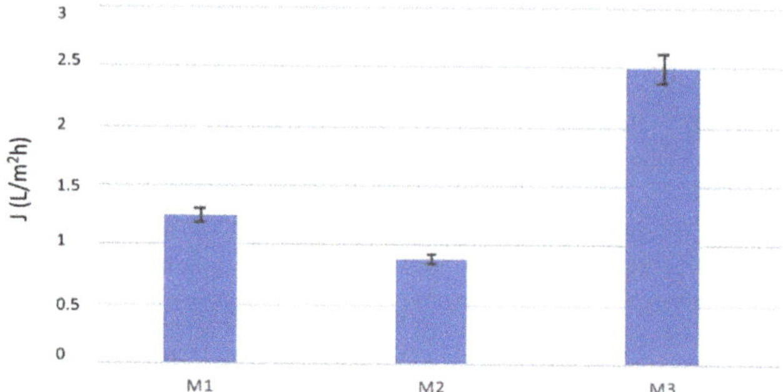

Figure 3. Transmembrane flux obtained with the three modules. v_{fin}, 0.35 m/s; v_{din}, 0.16 m/s; T_{fin}, 37 °C; T_{din}, 22 °C.

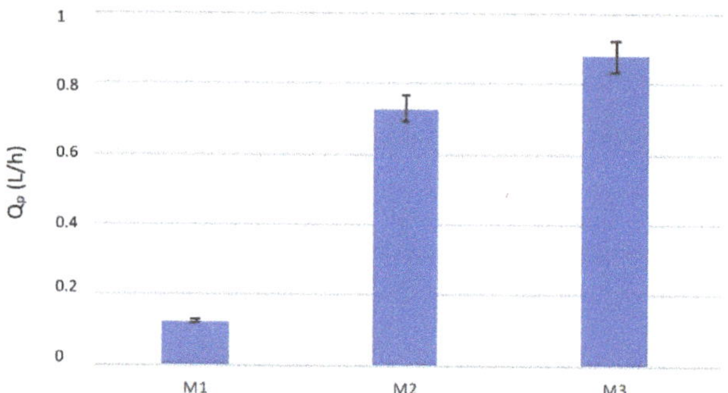

Figure 4. Permeate produced with the three modules. v_{fin}, 0.35 m/s; v_{din}, 0.16 m/s; T_{fin}, 37 °C; T_{din}, 22 °C.

The highest flux was achieved with the M3 module, followed by the M1, and the M2 module leading to the lowest value. Since the two Microdyn modules were equipped with the same type of membrane, a similar flux was expected if working at the same inlet conditions. However, a 29% lower flux was registered. This result can be due to the longer fibers of the M2 module (the M2 length was 75 cm vs. 50 cm of the M1), as well as to its not uniform packing of capillaries. When working with longer membranes, the feed and permeate temperatures along the module decrease and increase, respectively, more significanlty than with shorter membranes, and therefore, a lower driving force is available for the vapor transport. Furthermore, the worse result of the M2 module could also be attributed to a bad flow distribution of the permeate stream at the shell side, because of the bad packing of the capillaries that caused a non-uniform flow along the module and, possibly, a reduction of the effective membrane area available for the permeation where adjacent capillaries touch each other. If part of the outer membrane surface is not in contact with the permeate stream or if the permeate stationates/slowly flows inside empty spaces present among membranes, the vapor coming from the feed side is not efficiently condensed and removed, so that the resistance to the vapor transport is increased and heat accumulates at the shell side, with a consequent reduction in the driving force available. To support this hypothesis, the distribution of the fibers inside the two modules was inspected.

In Figure 5, the pictures of the two ends of each module are shown. It is evident that the capillaries were well and evenly packed in the smaller module, whilst in the bigger module, they were not uniformly distributed. In fact, some of them were very close and some others were quite far, leaving empty spaces. It is expected that the same happens inside the module, where it is difficult to ensure a fixed distance among capillaries.

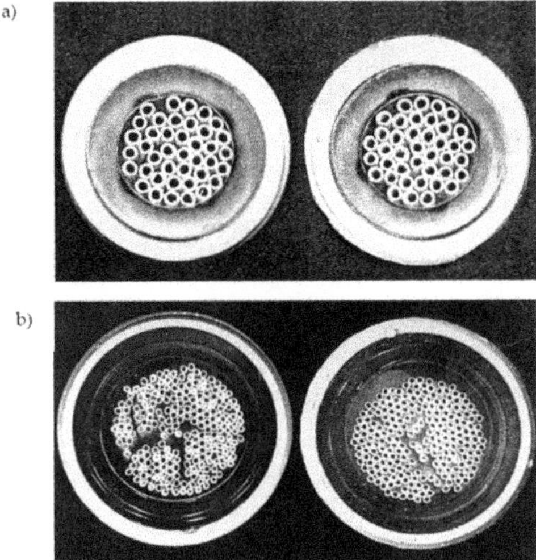

Figure 5. Picture of the (**a**) the M1 and (**b**) the M2 modules.

Therefore, with the two Microdyn modules, it was not possible to operate in a modular way: by working with a 8.3 higher membrane area, the productivity was increased by a factor of 5.9. The mean free path of the water vapor at 37 °C calculated by equation (2) was 0.1075 μm and the Knudsen coefficient varied between 0.36 and 0.54 for the M3 and the Microdyn membranes, respectively, indicating in both cases a transition region for the mass transport. The highest flux and productivity of the M3 module can be attributed to the different membrane properties, that presented a higher pore size and lower thickness than the capillary membranes used in the Microdyn modules. It has to be pointed out that the M3 module is also the shortest one, which is a favorable feature to reduce the feed temperature decay along the fibers. The three modules were also compared in terms of STEC. In particular, Figure 6 shows the STEC for the three modules operating at the same conditions. The M3 module led to the lowest STEC, due to the highest permeate production, while the two Microdyn modules had similar STEC values, although that of the M1 module was slightly lower. The STEC data are an indication of the energy efficiency of the modules, as they give an information of the thermal energy needed for a certain permeate production. Therefore, the lowest is the STEC, the highest is the energy efficiency of the unit. It has to be noticed that in lab-scale tests the STEC values are usually higher than those obtained with bigger membrane modules, due to the low membrane areas involved and the related low permeate produced. Nevertheless, in this work, due to the higher transmembrane fluxes, the M3 module (0.35 m^2) was able to produce more permeate than the M2 module (0.83 m^2), leading to their best performance. This result points out the importance of membrane properties and module features for improving the efficiency of membrane distillation units.

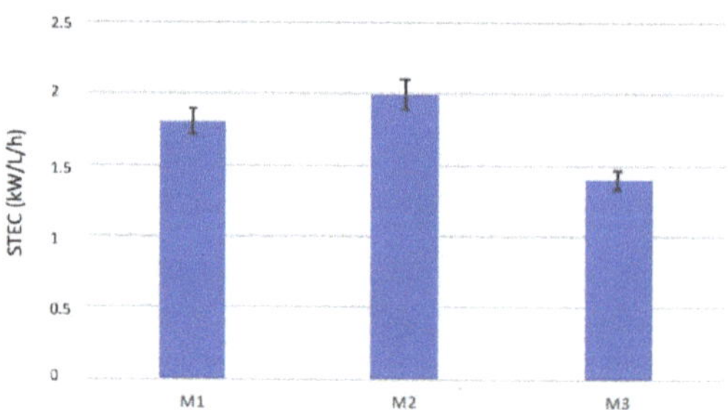

Figure 6. Specific thermal energy consumption (STEC) values for the three modules.

The size of the membrane distillation modules is also important, in the logic of the reduction of the land use and of the minimization of the occupied volume in space applications (process intensification strategy). Figure 7a compares the modules in terms of volume, showing that the two Microdyn modules had the highest (M1) and lowest (M2) volumes, with the M3 volume closer to the M1 one. In Figure 7b, the compactness (calculated as the membrane area divided by the volume) is reported for the three modules. The M2 was the less compact, followed by the M1 and finally by the M3 module, that was the most compact one. This result confirms the benefit of using smaller capillaries that allow to achieve high membrane area values in lower volumes.

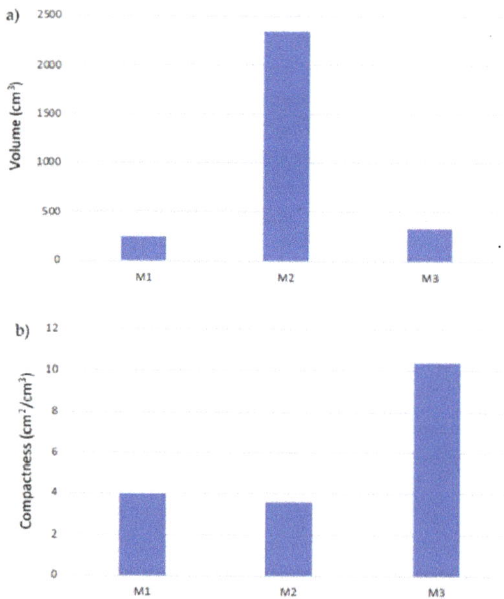

Figure 7. (**a**) Size (volume) and (**b**) compactness of the three modules.

Based on the obtained data, the three modules were finally compared in terms of PS, see Figure 8. The PS was calculated according to Equation (4), with the permeate flow rate Qp as the productivity and the volume of the module as size.

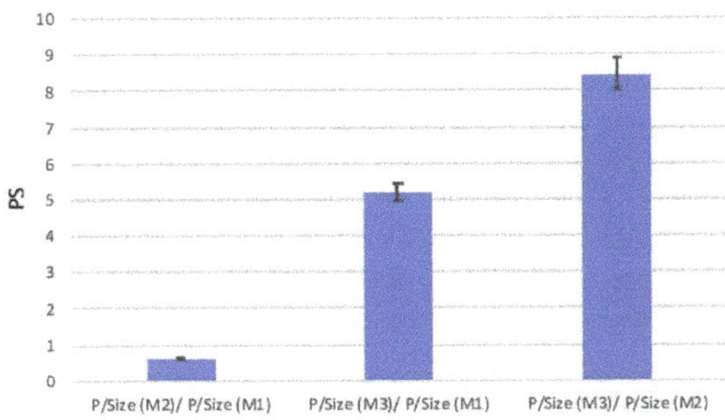

Figure 8. Modules' comparison in terms of the productivity/size ratio (PS).

With respect to the M1 module, the M2 presented a lower productivity per size (PS < 1), while the M3 module showed the best performance, leading to ratios much higher than 1 if compared with both Microdyn modules.

4. Conclusions

The potential of DCMD for treating at low temperature solutions containing urea was explored in few studies and therefore, the state of the art on this topic is still limited. Nevertheless, the researches made in the field confirmed the high rejections for urea of DCMD, that can be used to produce purified water together with a concentrate source for urea/nutrients recovery. This result is interesting for applying/improving the recycle and recovery concept in space engineering, in medical treatments, as well as in urea wastewater treatment plants.

Concerning the case study on the use of DCMD for recovery water from the human plasma ultrafiltrate, the 0.35 m^2 (M3) module led to the highest permeate production (0.88 L/h) and to the lowest STEC (1.4 kW/L/h), resulting also to be the most compact and with the lowest size. However, the productivity target is still far and therefore, there is the need of developing capillary membranes and modules "ad hoc", in order to obtain the desired distillate production.

From the carried out investigation, the difficulty in scaling-up capillary membrane modules, to ensure modularity, is evident. Nevertheless, some inputs for addressing membrane and module developments can be derived: specific attention should be made to the optimal size of the membranes, because small diameters allow a high compactness and low size of the module, but could limit the operating flow rates, due to the pressure drops, with a consequent increase in polarization phenomena and flux reduction. The permeate production can be also enhanced by using membranes with higher porosity and pore size and lower thickness, taking care of the wetting risk. Referring to the module development, shorter modules should be preferred, in order to reduce the feed temperature decay and to avoid capillary deformation. A fundamental step is to also ensure an uniform packing of the membranes inside the module, so that to have a constant cross section for the fluid flow and to effectively utilize all the membrane area for the distillation process.

Author Contributions: Conceptualization: A.C. (Alessandra Criscuoli), A.C. (Alfredo Capuano), M.A. and E.D.; data curation: A.C. (Alessandra Criscuoli); investigation: A.C. (Alessandra Criscuoli); supervision: E.D.; writing—original draft preparation: A.C. (Alessandra Criscuoli); writing—review and editing: A.C. (Alessandra Criscuoli). All authors have read and agreed to the published version of the manuscript.

Funding: This research received no external funding.

Conflicts of Interest: The authors declare no conflict of interest.

References

1. Ashoor, B.B.; Mansour, S.; Giwa, A.; Dufour, V.; Hasan, S.W. Principles and applications of direct contact membrane distillation (DCMD): A comprehensive review. *Desalination* **2016**, *398*, 222–246. [CrossRef]
2. Cath, T.Y.; Adams, D.; Childress, A.E. Membrane contactor processes for wastewater reclamation in space II. Combined direct osmosis, osmotic distillation, and membrane distillation for treatment of metabolic wastewater. *J. Membr. Sci.* **2005**, *257*, 111–119. [CrossRef]
3. Capuano, A.; Memoli, B.; Andreucci, V.E.; Criscuoli, A.; Drioli, E. Membrane distillation of human plasma ultrafiltrate and its theoretical applications to heamodialysis techniques. *Int. J. Art. Organs* **2000**, *23*, 415–422. [CrossRef]
4. Cartinella, J.L.; Cath, T.Y.; Flynn, M.T.; Miller, G.C.; Hunter, K.W., Jr.; Childress, A.E. Removal of natural steroid hormones from wastewater using membrane contactor processes. *Environ. Sci. Technol.* **2006**, *40*, 7381–7386. [PubMed]
5. Liu, Q.; Liu, C.; Zhao, L.; Ma, W.; Liu, H.; Ma, J. Integrated forward osmosis-membrane distillation process for human urine treatment. *Water Res.* **2016**, *91*, 45–54. [CrossRef] [PubMed]
6. Volpin, F.; Checkli, L.; Phuntsho, S.; Ghaffour, N.; Vrouwenvelder, J.S.; Shon, H.K. Optimisation of a forward osmosis and membrane distillation hybrid system for the treatment of source-separated urine. *Sep. Pur. Technol.* **2019**, *212*, 368–375. [CrossRef]
7. Khumalo, N.; Nthunya, L.; Derese, S.; Motsa, M.; Verliefde, A.; Kuvarega, A.; Mamba, B.B.; Mhlanga, S.; Dlamini, D.S. Water recovery from hydrolysed human urine samples via direct contact membrane distillation using PVDF/PTFE membrane. *Sep. Pur. Technol.* **2019**, *211*, 610–617. [CrossRef]
8. Camacho, L.M.; Dumée, L.; Zhang, J.; Duke, M.; Gomez, J.; Gray, S. Advances in Membrane Distillation for Water Desalination and Purification Applications. *Water* **2013**, *5*, 94–196. [CrossRef]
9. Criscuoli, A.; Carnevale, M.C.; Drioli, E. Evaluation of energy requirements in membrane distillation. *Chem. Eng. Proc. Proc. Intens.* **2008**, *47*, 1098–1105. [CrossRef]
10. Criscuoli, A.; Drioli, E. New metrics for evaluating the performance of membrane operations in the logic of process intensification. *Ind. Eng. Chem. Res.* **2007**, *46*, 2268–2271. [CrossRef]

Article

Treatment of Flue Gas Desulfurization Wastewater by an Integrated Membrane-Based Process for Approaching Zero Liquid Discharge

Carmela Conidi [1], Francesca Macedonio [1,2,*], Aamer Ali [1], Alfredo Cassano [1,*], Alessandra Criscuoli [1], Pietro Argurio [2] and Enrico Drioli [1]

1 Institute on Membrane Technology, ITM-CNR, c/o University of Calabria, via P. Bucci, 17/C, I-87036 Rende (CS), Italy; c.conidi@itm.cnr.it (C.C.); amir_hmmad@hotmail.com (A.A.); a.criscuoli@itm.cnr.it (A.C.); e.drioli@itm.cnr.it (E.D.)

2 Department of Environmental and Chemical Engineering, University of Calabria, Via P. Bucci, 44/A, I-87036 Rende (CS), Italy; pietro.argurio@unical.it

* Correspondence: francesca.macedonio@unical.it (F.M.); a.cassano@itm.cnr.it (A.C.); Tel.: +39-0984-492-012 (F.M.); +39-0984-492067(A.C.)

Received: 25 October 2018; Accepted: 22 November 2018; Published: 26 November 2018

Abstract: An integrated membrane process for the treatment of wastewaters from a flue gas desulfurization (FGD) plant was implemented on a laboratory scale to reduce their salt content and to produce a water stream to be recycled in the power industry. The process is based on a preliminary pretreatment of FGD wastewaters, which includes chemical softening and ultrafiltration (UF) to remove Ca^{2+} and Mg^{2+} ions as well as organic compounds. The pretreated wastewaters were submitted to a reverse osmosis (RO) step to separate salts from water. The RO retentate was finally submitted to a membrane distillation (MD) step to extract more water, thus increasing the total water recovery factor while producing a high-purity permeate stream. The performance of RO and MD membranes was evaluated by calculating salts rejection, permeate flux, fouling index, and water recovery. The investigated integrated system allowed a total recovery factor of about 94% to be reached, with a consequent reduction of the volume of FGD wastewater to be disposed, and an MD permeate stream with an electrical conductivity of 80 µS/cm, able to be reused in the power plant, with a saving in fresh water demand.

Keywords: FGD wastewater; integrated membrane-based process; zero liquid-discharge; sustainability

1. Introduction

Much research is today focusing on minimizing the scarcity of potable water and the impact of air, water, and solid waste pollutants. Among them, to improve the control of SO_2 emissions, various flue gas desulfurization (FGD) technologies have been developed in the last decades.

Flue gas desulfurization processes are primarily used to remove SO_2 from exhaust flue gases of fossil fuel thermoelectric power plants. In these plants, SO_2 is produced during the combustion of coal and oil, and can be further converted (about 1%) into sulfur trioxide (SO_3) if high contents of oxygen are present [1]. The commonly used FGD technologies are: Wet and spray-dry scrubbing (using a slurry of alkaline sorbents, like limestone or lime, or seawater); the wet sulfuric acid process (recovering sulfur as sulfuric acid); and dry sorbent injection systems.

Depending on the coal source, used technology, and operating conditions, FGD processes can give origin to various streams, with a different and complex composition. Chloride, sulfate, nitrate, calcium, magnesium as well as various heavy metals and dissolved silica and borate are often present. Moreover, FGD wastewater may have a content of total dissolved solids (TDS) as high as 50,000 mg/L [2].

The treatment of FGD wastewater represents one big issue in the water industry [2,3]. One possibility is to use biotechnological treatments, which, however, lead to the production of H_2S [4].

Another option consists in applying chemical precipitation-based strategies for heavy metal removal, which use alkaline compounds, such as $Ca(OH)_2$ (hydrated lime), NaOH (caustic soda), or Na_2CO_3 (soda ash), leading to the production of insoluble hydroxides. This approach is limited by the production of huge quantities of heavy metal hydroxide and calcium sulfate sludge that have to be disposed in a regulated hazardous waste facility.

The use of zero-valent iron (ZVI or Fe0) as reactive media for treating heavy metal contaminated groundwater has also been investigated in the last years [5–8]. The addition of iron promotes the removal of dissolved heavy metals by several mechanisms, including cementation, precipitation of metal hydroxides, and adsorption [9]. This approach has also been evaluated for the removal of selenium [10,11], mercury, and other heavy metals [12] from FGD wastewaters. In particular, continuous-flow field tests conducted on a fluidized bed system consistently reduced Hg from ca. 50 to <0.005 μg/L and Se (mostly as selenate) from ca. 3000 to <7 μg/L. Most of the heavy metals were all reduced to near- or sub-ppb levels [12]. However, the potential of ZVI as a reagent for remediating contaminated FGD wastewaters is limited by the rapid loss of ZVI reactivity upon the formation of iron corrosion products as a passive coating on the ZVI grains [13].

In recent years, the potential of membrane operations, like microfiltration (MF), ultrafiltration (UF), nanofiltration (NF), and reverse osmosis (RO), for treatment of different wastewaters has been investigated. The attraction of membrane operation was, among others, due to their capability of removing almost all pollutants with a reduced addition of chemicals (only the amount necessary for membrane pre-treatment and cleaning).

Apart from chemical and biotechnological processes, membrane operations can be a viable approach for the remediation of FGD wastewaters, although few studies have been reported until now on this subject. The performance of a coprecipitation method of heavy metal hydroxides and sulphides followed by crossflow microfiltration (CMF) in the treatment of wastewater from a FGD plant was analyzed by Enoch et al. [14]. The removal efficiency of both hydrophilic and hydrophobic membranes was satisfactory, except for Cd removal.

Liu et al. [15] also used MF membranes, with and without an initial chemical precipitation, for the removal of Hg^{2+} from FGD wastewater, demonstrating the feasibility of the process.

An integrated membrane system, constituted of a sedimentation tank for particles sedimentation, UF, NF, and RO, was utilized by Yin et al. [16] for the treatment of the desulfurization wastewater. In particular, the sedimentation step revealed to be fundamental to improve UF flux and permeate quality. As a matter of fact, without pre-sedimentation, a lower steady flux of 200 L/m^2h was reached, compared to about 500 L/m^2h achieved in the UF process with pre-sedimentation, at a transmembrane pressure (TMP) of 0.15 MPa, temperature of 40 °C, and cross-flow velocity (CFV) of 4 m/s. UF retained 99.9% of the initial suspended sulfur (SS) while salt compounds passed through the UF membrane and were recovered in the permeate. Therefore, UF ceramic membranes were considered for removing SS, generating a permeate stream able to be treated by the NF unit. The NF unit separated bivalent ammonium salts from monovalent ammonium salts and the obtained NF permeate was sent to RO to separate monovalent salts from water, and to recover ammonium thiocyanate. The analyzed integrated membrane process (UF/NF/RO) was able to reduce environmental loads and to make possible the recovery of various valuable substances (such as, sulfur and NH_4SCN) and the water reuse.

Being a pressure driven membrane operation that is well consolidated also at the industrial level, RO has the great advantage of producing high quality water (rejection factor higher than 99% for monovalent and bivalent ions in the feed) at a relatively low energy consumption (3 and 4 kWh/m^3 are typical values at around a 50% recovery factor). The main RO limitation is the high feed pressure required for overcoming the feed side osmotic pressure, which usually ranges from 55 to 68 bar for a salinity of about 35g/L, and that is around 15 bar for salinity between 0.5 and 30 g/L [17]. Being a thermally driven membrane separation technology, in membrane distillation (MD), operating pressures are generally on the order of a few hundred kPa. Moreover, its performance is less affected by concentration polarization than pressure driven membrane operations, allowing the treatment of higher salinity water and leading to a higher recovery factor, then reducing the environmental impact of the produced brine and with the potential to achieve near zero discharge. In fact, various studies proved that MD could accomplish nearly complete salt rejection and high water reclamation in the treatment of high-salinity wastewater [18–21], as is the case of FGD wastewater produced by seawater scrubbing. Moreover, MD has the additional advantage of a relatively low operating temperature, with the possibility to utilize either waste heat or renewable energy resources (such as geothermal or solar energy) or the low-grade heat available in power plants [22–24].

Jia and Wang [25] recently investigated an integrated process for the treatment of flue gas desulfurization wastewater based on chemical softening followed by NF and MD working on the NF permeate. They found that the chemical softening and the NF pretreatment could significantly decrease membrane scaling in MD. Moreover, over 99.99% salt rejection and over 92% of water reclamation were achieved. The MD configuration used by authors was the vacuum membrane distillation (VMD) because of its high flux. Nevertheless, this configuration needs the use of a vacuum pump and of an external condenser for the permeate recovery.

In a previous work, we evaluated the performance of two commercial RO spiral-wound membranes (SWC-2540 and ESPA-2540 both from Hydranautics) for the removal of salt compounds from softened and ultrafiltered FGD wastewaters [26]. The SWC membrane was more effective than the ESPA membrane in terms of ions rejection, fouling index, and cleaning efficiency, and a permeate conductivity lower than 2 mS/cm was obtained. Based on the positive results achieved, in the current study, we investigated the possibility to treat the RO brine produced by a SWC-2540 RO membrane in an MD unit, with the aim of producing higher purity water as MD permeate, while reducing the brine volume to be disposed. As in the previous study [26], FGD wastewaters were submitted to a pre-treatment step (softening and UF) aimed at reducing Ca^{2+} and Mg^{2+} content, as well as the organic content, to minimize scaling. The UF permeate was then processed by the SWC-2540 RO membrane and its brine was finally submitted to an MD step (Figure 1). Direct Contact Membrane Distillation (DCMD) was chosen because, among the MD configurations, it is the simplest to operate, requiring the least equipment.

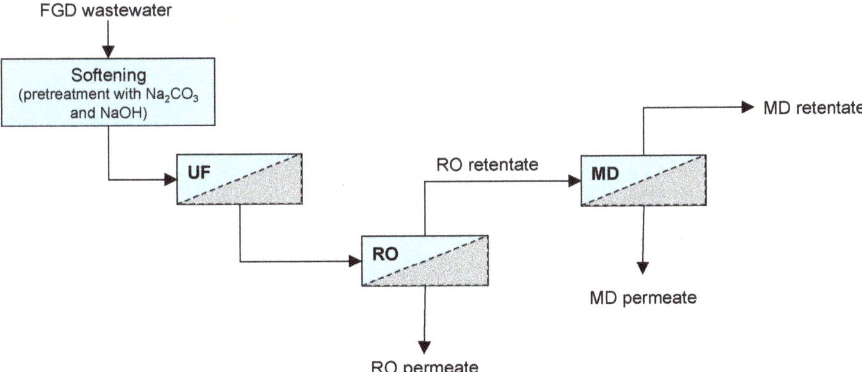

Figure 1. Scheme of the integrated membrane system investigated.

2. Materials and Methods

2.1. Pretreatment of FGD Wastewaters

FGD wastewaters were collected from the thermal power plant of Industrial Enel Research (Brindisi, Italy). The whole pretreatment process consisted of softening agents' addition followed by UF.

FGD wastewaters were softened with $Na_2CO_3{\cdot}H_2O$ ($Na_2CO_3{\cdot}H_2O$/Ca molar ratio 2.6/1) and NaOH (NaOH/Mg molar ratio 1.5/1); the solutions were incubated for 1h at room temperature and then pre-filtered with filter paper (pore size of about 10–20 μm) before acidification with H_2SO_4 up to pH 6.5. Afterwards, the FGD wastewaters were treated with an antiscalant (Carboxyline CM supplied by Aquastill B.V., Sittard, The Netherlands) at the recommended concentration (8 mg/L) in order to prevent the precipitation of low soluble salts.

Cross-flow UF was performed by using a hollow fiber polyethersulphone (PES) membrane module (FUS 5082, Microdyn Nadir, Wiesbaden, Germany), having a nominal molecular weight cut-off (MWCO) of 500 kDa and an effective membrane area of 0.25 m^2. A feed-and-bleed configuration (in order to work at constant feed volume, during the experiments feed was added in the feed tank in the same amount of the produced permeate) was operated until a recovery factor (RF) of 98% was obtained at the following experimental conditions: TMP of 0.45 bar, axial feed flowrate (Q_f) of 560 L/h, and temperature (T) of 25 ± 1 °C. In the selected operating conditions, average permeate fluxes in the range of 460–500 kg/m^2h were registered.

After the treatment with FGD wastewater, the UF membrane module was rinsed with distilled water for 30 min and then cleaned by recycling an acid detergent (Ultraclean WO, Henkel Chemicals Ltd., Dusseldorf) at a concentration of 0.1% (w/w) (pH 4) for 60 min at 40 °C. After, the membrane module was rinsed with distilled water for 20 min. The cleaning-in-place procedure allowed more than 98% of the initial permeability of the UF membrane to be recovered.

2.2. RO and MD: Experimental Set-Up and Membranes

RO experiments were carried out using the laboratory set-up of Matrix Desalination Inc. (Florida, USA), which incorporated a stainless steel housing able to accommodate a 2.4 × 21 in. spiral-wound membrane module with an effective membrane surface area of 2.34 m^2 (main properties are listed in Table 1). The equipment consists of a feed tank with a capacity of 20 L, a high-pressure pump, a back-pressure valve, two pressure gauges, a permeate control valve, and a coiling cool fed with tap water used to maintain the control the feed temperature.

Table 1. Characteristics of the reverse osmosis (RO) membrane module.

Membrane Type	SWC-2540
Membrane Material	Composite polyamide
Configuration	Spiral-wound
Salt Rejection (%)	99.4 (minimum 99.0)
pH Operating Range	2–11
Max. Operating Temperature (°C)	45
Max. Operating Pressure (bar)	69
Membrane Surface Area (m^2)	2.34
Water Permeability (kg/m^2hbar)	1.77 [a]
Zeta Potential (mV)	−21.2 at pH 7 [b]
Contact Angle	96.05 ± 4.35 [c]

[a] our data; [b] data from Li et al. [27]; [c] data from Lee et al. [28].

The pretreated FGD wastewaters were treated in a spiral-wound RO membrane module (SWC-2540) supplied by Hydranautics Corporation (Oceanside, CA, USA), whose properties are summarized in Table 1. Experiments were performed according to a feed-and-bleed configuration in selected operating conditions (TMP, 36 bar; Q_f, 204 L/h; T, 26.5 °C) up to an RF of 60%.

The permeate flux was determined by weighing the amount of permeate collected vs. time and using the following equation:

$$J_p = \frac{m_p}{A \cdot t} \quad (1)$$

where J_p is the permeate flux (kg/m^{2}h), m_p the permeate weight (kg) at time, t (h), and A is the membrane surface area (m^2).

The water permeability (W_p) of the membrane was obtained as the slope of the straight line resulting from plotting the water flux at 25 °C versus the applied TMP.

After each experiment, the RO membrane was rinsed with water at 30 °C and then cleaned with an acid solution (Ultraclean WO 0.05%, pH 4) at 40 °C, for 60 min. Then, the membrane was rinsed with distilled water for 20 min and the water permeability was remeasured.

The fouling index of the RO membrane (FI_{RO}) was calculated by the following equation:

$$FI_{RO} = (1 - \frac{W_{p1}}{W_{p0}}) \cdot 100 \quad (2)$$

where W_{p0} and W_{p1} are the water permeability measured before and after the FGD wastewater treatment.

The cleaning efficiency was evaluated by using the water flux recovery method, according to the following equation:

$$CE = (\frac{W_{p3}}{W_{p0}}) \cdot 100 \quad (3)$$

where W_{p3} is the water permeability measured after the chemical cleaning.

Membrane distillation was performed in a direct contact configuration (DCMD). In the lab plant, the MD feed (i.e., RO retentate) and the MD permeate streams (i.e., demineralized water) converge in the counter-current mode towards the membrane module containing the flat oleophobic membrane supplied by Aquastill. The main properties of the used membrane are summarized in Table 2.

Table 2. Characteristics of the oleophobic membrane used in membrane distillation (MD).

Membrane Material	Polyethyelene-oleophobic (PE-O)
Configuration	Flat sheet
Active Module Length	50 cm
Membrane Area	0.05 m^2
Mean Pore Size	0.3 µm
Porosity	80%
Membrane Thickness	76 µm
Liquid Entry Pressure (LEP)	>4 bar
Contact Angle	>118°

The driving force in DCMD is a vapour pressure difference across the membrane, which is imposed by a temperature difference across the membrane. Therefore, the retentate line was heated by an ISCO GTR 2000 heater (Isco srl, Fizzonasco di Pieve Emanuele (MI), Italy) whilst the permeate line was cooled by a RTE 17 NESLAB refrigerated bath chiller circulator (Thermo Electron Corporation, Newington, CT, USA). The retentate solution and distillate coming out from the module were returned back to the feed tank and permeate tank, respectively, both working at atmospheric pressure. Magnetic drive gear pumps (Iwaki Co., Ltd., Tokyo, Japan) were used to recirculate the streams. The plant was also equipped with flow meters, thermocouples, manometers, and a conductivity meter.

The trans-membrane flux was calculated by evaluating weight variations in the distillate tank by a Gibertini EU-C LCD balance (Gibertini Elettronica, Novate Milanese (MI), Italy). After each experiment, the MD membrane was rinsed with water. The fouling index of the MD membrane (FI_{MD}) was calculated by:

$$FI_{MD} = \left(1 - \frac{J_{w1}}{J_{w0}}\right) \cdot 100 \quad (4)$$

where J_{w0} and J_{w1} are the MD trans-membrane flux before and after RO retentate treatment, respectively, when distilled water is used as the MD feed and the membrane process is carried out at 48 °C.

On the basis of previous MD experiments directly performed on FGD wastewaters [29], experimental runs were carried out at the operative conditions reported in Table 3.

Table 3. MD operative conditions.

MD Feed Solution	RO brine
$T_{Feed, in}$, °C	69 ± 0.1
$T_{Permeate, in}$, °C	28 ± 0.3
Feed Flow Rate, l/min	0.5
Permeate Flow Rate, l/min	0.4

2.3. Analytical Measurements

The different samples collected from the investigated processes were analyzed for electrical conductivity (EC), total dissolved solids (TDS), Ca^{2+}, Mg^{2+}, Na^{+}, TOC, and pH.

The removal efficiency for each component was expressed in terms of rejection (R) according to the following equation:

$$R = 1 - \left(\frac{C_p}{C_f}\right) \cdot 100 \quad (5)$$

where C_f and C_p are the concentrations of a specific component in the feed and permeate, respectively.

EC and TDS were measured using a digital conductivity meter (HI 2300 Microprocessor Conductivity, Hanna Instruments, Woonsocket, RI, USA).

Ca^{2+}, Mg^{2+}, and Na^{+} concentrations were determined by using a high-resolution continuum source atomic absorption spectrometer (HR-CSAAS, ContrAA700, Analytik Jena AG, Jena, Germany), with a high intensity Xe short arc lamp as a continuum source. Samples and standards were appropriately diluted (300 times for Mg and Ca, 3000 times for Na). Subsequently, they were acidified with 1% HCl and the absorbance measurements were performed using the spectral lines at 422.67 nm, 588.99 nm, and 285.21 nm for Ca^{2+}, Na^{+}, and Mg^{2+}, respectively.

pH was measured by an Orion Expandable ion analyzer EA 920 pH meter (Allometrics, Inc., Baton Rouge, LA, USA) with automatic temperature compensation.

Total organic carbon (TOC) was analyzed by a TOC analyzer (TOC-V CSN, Shimadzu, Kyoto, Japan).

3. Results and Discussion

3.1. Pretreatment of FGD Wastewaters

The chemical composition of FGD wastewater before and after the pre-treatment step is reported in Table 4. Raw waters were characterized by a lower content of Ca^{2+} and Mg^{2+} when compared to typical FGD wastewaters sampled by United States Environmental Protection Agency (USEPA) in several US thermal plants (Ca^{2+} in the range of 2000–5400 ppm and Mg^{2+} in the range of 1000–4200 ppm, respectively) [3]. Indeed, the concentration of Ca^{2+} and Mg^{2+} before the pre-treatment was approximately 384.4 ± 4.8 ppm and 289.9 ± 2.6 ppm, respectively. On the other hand, the average content of Na^{+} in the raw water was significantly higher (7.28 ± 0.6 g/L) than that found in typical wastewaters sampled by USEPA (50–2000 mg/L) [3]. These differences can be attributed to several aspects, which contribute to the pollutant concentrations of FGD wastewaters, including the coal type, the sorbent used, the materials of construction in the FGD system, and the FGD system operation. The addition of sodium carbonate and sodium hydroxide in the conditions previously optimized [26] permitted the content of Ca^{2+} and Mg^{2+} of raw wastewaters to be reduced, with a removal efficiency of 90.0% and 78.8%, respectively. A lower removal efficiency was measured for Na+ (3.84%). The chemical pre-treatment allowed the achievement of a 7.4% removal of TDS (from 16.9 ± 0.6 g/L to 15.7 ± 0.8 g/L);

a further removal was reached after the UF treatment, with a final overall value of 13.49%. The UF process slightly changed the concentrations of the analysed compounds if compared with the chemical pre-treated FGD wastewater, as it could be expected from the MWCO of the used UF membrane. Nevertheless, the UF step removed more than 60% of TOC. This rejection value can be explained assuming the presence of not totally dissolved organic solids forming micro-droplets with a size in the range of the MWCO of the UF membrane.

Table 4. Chemical composition of flue gas desulfurization (FGD) wastewater before and after pre-treatment.

Parameter	Sample			Overall Removal (%)
	Raw Water	After Softening	After UF	
Ca^{2+} (ppm)	384.4 ± 4.8	39.16 ± 2.1	38.13 ± 2.1	90.00
Mg^{2+} (ppm)	289.9 ± 2.6	62.5 ± 0.5	62.4 ± 0.1	78.84
Na^{+} (g/L)	7.28 ± 0.6	7.0 ± 0.14	7.0 ± 0.3	3.84
EC (mS/cm)	33.6 ± 2.1	32.5 ± 1.2	31.1 ± 1.7	7.44
TDS (g/L)	16.9 ± 0.6	15.7 ± 0.8	14.62 ± 1.2	13.49
TOC (mg/L)	-	90.12 ± 0.90	33.80 ± 0.34	62.50
pH	6.7 ± 0.1	6.55 ± 0.2	6.8 ± 0.1	-

3.2. Reverse Osmosis and Membrane Distillation Performance

The RO membrane performance was assessed in terms of the permeate flux (J_p), membrane cleaning efficiency, and salt removal efficiencies. The time evolution of the permeate flux of pre-treated wastewaters in selected operating conditions is illustrated in Figure 2. Experimental data are referred to the processing of 16.1 kg of pre-treated FGD wastewaters with a production of 10.1 kg of permeate (recovery of about 63%). The initial permeate flux of 11 kg/m^{2}h sharply decreased during the first 20 min, then continued to reduce in time, reaching a pseudo-steady state value higher than 1 kg/m^{2}h after 120 min. The permeate flux decline can be due to different factors, like the increase of the osmotic pressure during continuous concentration of the feed solution, as well as to membrane fouling and concentration polarization phenomena [30]. In particular, concentration polarization leads to a higher solute concentration at the membrane surface, which may cause salts deposition and also high osmotic pressure at the membrane surface, which means a flux decline when RO is performed at constant pressure [31].

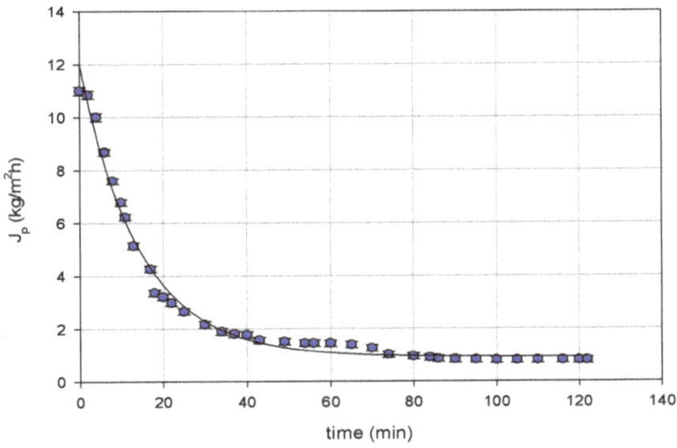

Figure 2. Reverse osmosis (RO) of pre-treated FGD wastewaters. Time course of permeate flux. (TMP = 36 bar; T = 26.5 °C; Q_f = 204 L/h).

Figure 3 shows the hydraulic permeability of the RO membrane before and after the treatment of the UF permeate and after cleaning procedures: The initial hydraulic permeability (W_{p0}) of about 1.73 kg/m^{2}hbar was reduced to 0.81 kg/m^{2}hbar after the treatment of the UF permeate. According to these data, the fouling index was about 53.1%. A first cleaning with distilled water at 30 °C recovered about 86% of the initial permeability (W_{p2} = 1.49 kg/m^{2}hbar); a higher water flux recovery (of about 96%) was reached through the chemical cleaning with acid detergent (W_{p3} = 1.49 kg/m^{2}hbar). Therefore, experimental data confirmed that acid solutions are effective in removing membrane scaling of RO membranes [32].

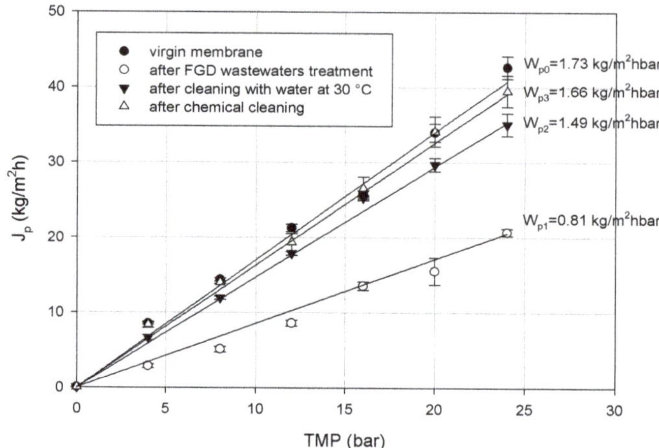

Figure 3. Permeate flux variation with TMP for the RO membrane before and after cleaning procedures (W_{p0} initial hydraulic permeability; W_{p1} hydraulic permeability after RO treatment; W_{p2} hydraulic permeability after cleaning with water; W_{p3} hydraulic permeability after cleaning with acid detergent).

The RO retentate was further concentrated by DCMD. The trend of the transmembrane flux (J_w) as a function of the MD recovery factor is reported in Figure 4. Experimental results indicated that the MD flux was not significantly affected by the recovery factor despite the growing feed concentration. The measured average flux was about 11 kg/m^{2}h. Additionally, the MD water recovery factor was also excellent and equal to 89.7%. Moreover, the quite constant trend of the trans-membrane flux indicated that no significant fouling occurred in the MD test. This was confirmed by the value of FI_{MD} (1.76%) determined via Equation (3).

These results were in agreement with those reported by Jia and Wang [25] in the treatment of FGD wastewaters by NF and MD. In their approach, the authors employed chemical softening and NF as pretreatment followed by VMD. The membrane flux of the VMD process remained stable during the whole continuous concentration and was of the order of 8.5 L/m^{2}h. On the other hand, the flux of a direct VMD process, which employed only MF as pretreatment, decreased sharply in the first 6h due to the formation of $CaSO_4$.

The chemical composition of the different samples coming from the RO/MD integrated process is summarized in Table 5. The EC of FGD wastewaters in the RO permeate was lowered down to 5.08 ± 0.10 mS/cm, with a removal efficiency of 85.4%.

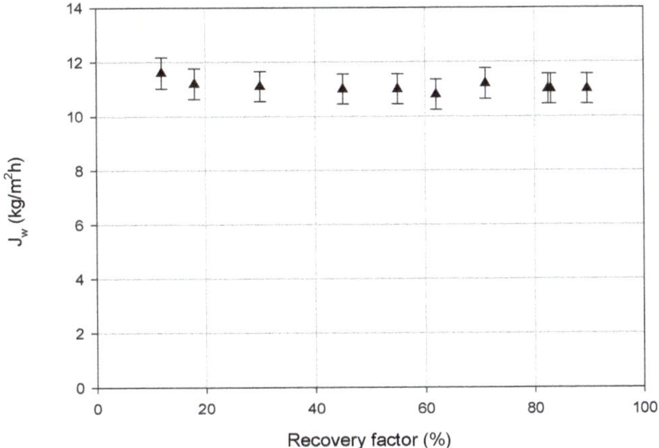

Figure 4. MD trans-membrane flux vs recovery factor.

Table 5. Chemical composition of FGD wastewaters coming from RO/MD treatments.

Sample	Ca^{2+} (ppm)	Mg^{2+} (ppm)	Na^{+} (g/L)	EC (mS/cm)	TDS (g/L)	pH
Feed RO	40.1 ± 0.8	67.4 ± 1.3	6.9 ± 0.1	34.7 ± 0.7	17.4 ± 0.9	7.53 ± 0.4
Permeate RO	6.9 ± 0.4	n.d.	0.70 ± 0.01	5.08 ± 0.10	2.53 ± 0.10	7.12 ± 0.14
Retentate RO	92.7 ± 1.8	175.1 ± 3.5	15.9 ± 0.3	69.4 ± 1.4	35.8 ± 0.7	7.78 ± 0.15
Permeate MD	16.22 ± 0.32	n.d.	n.d.	0.080 ± 0.001	0.040 ± 0.001	6.37 ± 0.13
Retentate MD	286.64 ± 5.37	539.8 ± 10.8	4.9 ± 0.1	158.3 ± 3.1	78.8 ± 1.6	8.15 ± 0.16

n.d.: not detectable.

TDS, Ca^{2+}, and Na^{+} ion concentrations in the RO permeate strongly reduced when compared with the feed solution. The observed rejection of the RO membrane towards Na^{+} was of about 90%, whereas, for Ca^{2+}, TDS, and EC, rejections were higher than 83%. Mg^{2+} ions were completely removed by the RO membrane, with a removal efficiency of 100% (Figure 5).

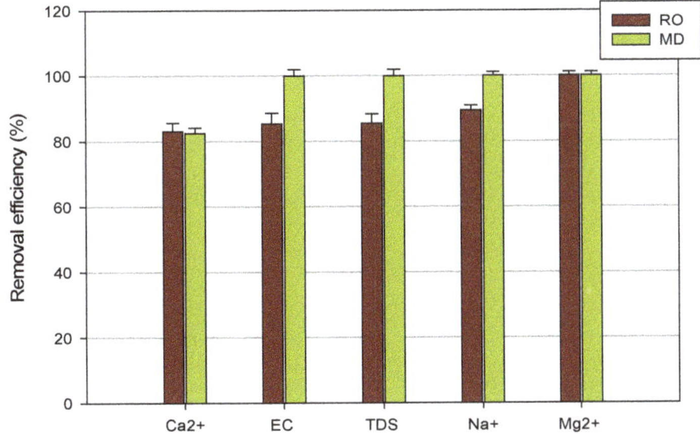

Figure 5. Removal efficiency of RO and MD membranes towards analyzed compounds.

The MD step removed both Mg^{2+} and Na^{+} ions from the RO retentate (Figure 5). TDS and EC rejections were also very high and equal to 99.9%. A lower retention value was measured for Ca^{2+}

ions (about 82.5%). This lower rejection could be attributed to calcium scaling on the membrane, which, however, did not affect the transmembrane flux (see Figure 4).

According to the experimental results, the combination of RO and MD processes reached a total recovery factor of about 94% and an MD permeate with a quality standard suitable to be reused in the power plant (needed purity: EC < 800 μS/cm), with a consequent saving in fresh water consumption and reduction of the volume of FGD wastewater to be disposed.

4. Conclusions

In the last decade, various flue gas desulfurization (FGD) technologies have been developed for removing sulfur dioxide from flue gases coming from fossil fuel thermoelectric power plants. FGD processes give origin to high salinity streams with a complex composition and the efficient treatment of FGD wastewater is one of the biggest challenges today.

In the present work, the problems of reducing water demand in power plants and of minimizing FGD wastewater discharge were dealt with. For this purpose, the potential of an integrated membrane-based process for FGD wastewater treatment and reuse was evaluated. In particular, the lab scale plant included: (1) A pre-treatment (chemical softening and UF) for reducing Ca^{+2}, Mg^{2+}, and TOC concentration in the raw wastewater, (2) an RO unit for the separation of salts from water, and (3) a MD unit for the treatment of RO retentate to extract more water, thus increasing the total water recovery factor of the process.

The experimental results indicated that high quality RO and MD permeate streams were obtained. In particular, the integrated RO/MD process achieved a total recovery factor of about 94% and an MD permeate stream with an electrical conductivity of 80 μS/cm that makes it suitable to be reused in the power plant. This will imply significant benefits in terms of a reduction of water demand in the plant, minimization of wastewater to be discharged in the environment, and overall improvement in the sustainability of the process.

Author Contributions: E.D. supervised the technical activities and the paper writing. Alfredo Cassano, Alessandra Criscuoli and F.M. conceived and designed the experiments (softening and membrane treatment of FGD wastewaters). Lab scale experiments were performed by C.C. and A.A. P.A. and C.C. were involved in the analytical characterization of raw and pretreated feed, permeate and retentate samples. All Authors contributed to the interpretation and discussion of experimental results.

Funding: This work was performed within the Project "Materials Technologies for performance improvement of Cooling Systems in Power Plants" (MATChING) which received funding from the European Union's Horizon 2020 research and innovation program under the grant agreement number 686031.

Conflicts of Interest: The authors declare no conflict of interest.

References

1. Cheremisinoff, N.P. *Pollution Control Handbook for Oil and Gas Engineering*; Scrivener Publishing LLC: Salem, MA, USA, 2016; pp. 503–511.
2. Chu, P. *Technical Manual: Guidance for Assessing Wastewater Impacts of FGD Scrubbers*; EPRI Report#1013313; Electric Power Research Institute: Palo Alto, CA, USA, 2006.
3. USEPA. *Steam Electric Power Generating Point Source Category: Final Detailed Study Report*; EPA 821-R-09-008; USEPA: Washington, DC, USA, 2009. Available online: https://www.epa.gov/sites/production/files/2015-06/documents/steam-electric_detailed_study_report_2009.pdf (accessed on 26 November 2018).
4. Mcvaugh, J.; Wall, W.T. Optimization of heavy metals wastewater treatment effluent quality versus sludge treatment. In Proceedings of the 31st Industrial Waste Conference, Purdue University, Lafayette, IN, USA, 22–26 March 1976; pp. 17–25.
5. Blowes, D.W.; Ptacek, C.J.; Benner, S.G.; McRae, C.W.T.; Bennett, T.A.; Puls, R.W. Treatment of inorganic contaminants using permeable reactive barriers. *J. Contam. Hydrol.* **2000**, *45*, 123–137. [CrossRef]
6. Morrison, S.J.; Metzler, D.R.; Dwyer, B.P. Removal of As, Mn, Mo, Se, U, V and Zn from groundwater by zero-valent iron in a passive treatment cell: Reaction progress modeling. *J. Contam. Hydrol.* **2002**, *56*, 99–116. [CrossRef]

7. Huang, Y.H.; Zhang, T.C.; Shea, P.J.; Comfort, S.D. Effects of oxide coating and selected cations on nitrate reduction by iron metal. *J. Environ. Qual.* **2003**, *32*, 1306–1315. [CrossRef] [PubMed]
8. Wilkin, R.T.; McNeil, M.S. Laboratory evaluation of zero-valent iron to treat water impacted by acid mine drainage. *Chemosphere* **2003**, *53*, 715–725. [CrossRef]
9. Shokes, T.E.; Moller, G. Removal of dissolved heavy metals from acid rock drainage using iron metal. *Environ. Sci. Technol.* **1999**, *33*, 282–287. [CrossRef]
10. EPRI. *Selenium Removal by Iron Cementation from a Coal-Fired Power Plant Flue-Gas Desulfurization Wastewater in a Continuous Flow System—A Pilot Study*; EPRI Report#1017956; Electric Power Research Institute: Palo Alto, CA, USA, 2009.
11. NAMC. *Review of Available Technologies for the Removal of Selenium from Water*; Technical Report; North America Metals Council: Washington, DC, USA, 2010; Available online: http://www.namc.org/docs/00062756.PDF (accessed on 10 July 2018).
12. Huang, Y.H.; Peddi, P.K.; Tang, C.; Zeng, H.; Teng, X. Hybrid zero-valent iron process for removing heavy metals and nitrate from flue-gas-desulfurization wastewater. *Sep. Purif. Technol.* **2013**, *118*, 690–698. [CrossRef]
13. Scherer, M.M.; Richter, S.; Valentine, R.L.; Alvarez, P.J. Chemistry and microbiology of permeable reactive barriers for in situ groundwater cleanup. *Crit. Rev. Environ. Sci. Technol.* **2000**, *30*, 363–411. [CrossRef]
14. Enoch, G.D.; Van Den Broeke, W.F.; Spiering, W. Removal of heavy metals and suspended solids from wastewater from wet lime (stone)-gypson flue gas desulphurization plants by means of hydrophobic and hydrophilic crossflow microfiltration membranes. *J. Membr. Sci.* **1994**, *87*, 191–198. [CrossRef]
15. Liu, C.; Farooq, K.; Doll, B.; Venkatadri, R. Economical and reliable mercury reduction in refinery and power plant discharge wastewater with robust microfiltration membrane technology. *Desalin. Water Treat.* **2013**, *51*, 4980–4986. [CrossRef]
16. Yin, N.; Liu, F.; Zhong, Z.; Xing, W. Integrated membrane process for the treatment of desulfurization wastewater. *Ind. Eng. Chem. Res.* **2010**, *49*, 3337–3341. [CrossRef]
17. Fritzmann, C.; Lowenberg, J.; Wintgens, T.; Melin, T. State-of-the-art of reverse osmosis desalination. *Desalination* **2007**, *216*, 1–76. [CrossRef]
18. Liu, H.Y.; Wang, J.L. Treatment of radioactive wastewater using direct contact membrane distillation *J. Hazard. Mater.* **2013**, *261*, 307–315. [CrossRef] [PubMed]
19. Kezia, K.; Lee, J.; Weeks, M.; Kentish, S. Direct contact membrane distillation for the concentration of saline dairy effluent. *Water Res.* **2015**, *81*, 167–177. [CrossRef] [PubMed]
20. Macedonio, F.; Ali, A.; Poerio, T.; El-Sayed, E.; Drioli, E.; Abdel-Jawad, M. Direct contact membrane distillation for treatment of oilfield produced water. *Sep. Purif. Technol.* **2014**, *126*, 69–81. [CrossRef]
21. Quist-Jensen, C.A.; Macedonio, F.; Horbez, D.; Drioli, E. Reclamation of sodium sulfate from industrial wastewater by using membrane distillation and membrane crystallization. *Desalination* **2017**, *401*, 112–119. [CrossRef]
22. Chafidz, A.; Kerme, E.D.; Wazeer, I.; Khalid, Y.; Ajbar, A.; Al-Zahrani, S.M. Design and fabrication of a portable and hybrid solar-powered membrane distillation system. *J. Clean. Prod.* **2016**, *133*, 631–647. [CrossRef]
23. Eykens, L.; Hitsov, I.; De Sitter, K.; Dotremont, C.; Pinoy, L.; Nopens, I.; Van Der Bruggen, B. Influence of membrane thickness and process conditions on direct contact membrane distillation at different salinities. *J. Membr. Sci.* **2016**, *498*, 353–364. [CrossRef]
24. Koschikowski, J.; Wieghaus, M.; Rommel, M. Solar thermal-driven desalination plants based on membrane distillation. *Desalination* **2003**, *156*, 295–304. [CrossRef]
25. Jia, D.; Wang, J. Treatment of flue gas desulfurization wastewater with near-zero liquid discharge by nanofiltration-membrane distillation process. *Sep. Sci. Technol.* **2018**, *53*, 146–153. [CrossRef]
26. Conidi, C.; Macedonio, F.; Argurio, P.; Cassano, A.; Drioli, E. Performance of reverse osmosis membranes in the treatment of flue-gas desulfurization (FGD) wastewaters. *Environments* **2018**, *5*, 71. [CrossRef]
27. Li, Q.; Song, J.; Yu, H.; Li, Z.K.; Pan, X.H.; Yang, B. Investigating the microstructures and surface features of seawater RO membranes and the dependencies of fouling resistance performances. *Desalination* **2014**, *352*, 109–117. [CrossRef]
28. Lee, S.; Lee, E.; Elimelech, M.; Hong, S. Membrane characterization by dynamic hysteresis: Measurements, mechanisms, and implications for membrane fouling. *J. Membr. Sci.* **2011**, *366*, 17–24. [CrossRef]

29. Ali, A.; Criscuoli, A.; Macedonio, F.; Argurio, P.; Figoli, A.; Drioli, E. Direct contact membrane distillation for treatment of wastewater for cooling tower in power industry. *H2 Open J.* **2018**, *1*, 57–68. [CrossRef]
30. Sim, L.N.; Chong, T.H.; Taheri, A.H.; Sim, S.T.V.; Lai, L.; Krantz, W.B.; Fane, A.G. A review of fouling indices and monitoring techniques for reverse osmosis. *Desalination* **2018**, *434*, 169–188. [CrossRef]
31. Shirazi, S.; Lin, C.J.; Chen, D. Inorganic fouling of pressure-driven membrane processes—A critical review. *Desalination* **2010**, *250*, 236–248. [CrossRef]
32. Filloux, E.; Wang, J.; Pidou, M.; Gernjak, W.; Yuan, Z. Biofouling and scaling control of reverse osmosis membrane using one-step cleaning-potential of acidified nitrite solution as an agent. *J. Membr. Sci.* **2015**, *495*, 276–283. [CrossRef]

Article

Concentration of 1,3-dimethyl-2-imidazolidinone in Aqueous Solutions by Sweeping Gas Membrane Distillation: From Bench to Industrial Scale

Ricardo Abejón [1,2], Hafedh Saidani [1], André Deratani [1], Christophe Richard [3] and José Sánchez-Marcano [1,*]

1 Institut Européen des Membranes UMR 5635, CNRS, ENSCM, Université de Montpellier, CC 047, Place Eugène Bataillon, 34095 Montpellier, France; ricardo.abejon@unican.es (R.A.); yahafedh@gmail.com (H.S.); andre.deratani@umontpellier.fr (A.D.)

2 Department of Chemical and Biomolecular Engineering, University of Cantabria, Avda. Los Castros s/n, 39005 Santander, Cantabria, Spain

3 Kermel, 20 Rue Ampère, CEDEX, 68027 Colmar, France; christophe.richard68@wanadoo.fr

* Correspondence: jose.sanchez-marcano@umontpellier.fr

Received: 6 November 2019; Accepted: 24 November 2019; Published: 26 November 2019

Abstract: Sweeping gas membrane distillation (SGMD) is a useful option for dehydration of aqueous solvent solutions. This study investigated the technical viability and competitiveness of the use of SGMD to concentrate aqueous solutions of 1,3-dimethyl-2-imidazolidinone (DMI), a dipolar aprotic solvent. The concentration from 30% to 50% of aqueous DMI solutions was attained in a bench installation with Liqui-Cel SuperPhobic® hollow-fiber membranes. The selected membranes resulted in low vapor flux (below 0.15 kg/h·m^2) but were also effective for minimization of DMI losses through the membranes, since these losses were maintained below 1% of the evaporated water flux. This fact implied that more than 99.2% of the DMI fed to the system was recovered in the produced concentrated solution. The influence of temperature and flowrate of the feed and sweep gas streams was analyzed to develop simple empirical models that represented the vapor permeation and DMI losses through the hollow-fiber membranes. The proposed models were successfully applied to the scaling-up of the process with a preliminary multi-objective optimization of the process based on the simultaneous minimization of the total membrane area, the heat requirement and the air consumption. Maximal feed temperature and air flowrate (and the corresponding high operation costs) were optimal conditions, but the excessive membrane area required implied an uncompetitive alternative for direct industrial application.

Keywords: sweeping gas membrane distillation; 1,3-dimethyl-2-imidazolidinone; solvent dehydration; hollow-fiber membrane; multi-objective optimization

1. Introduction

Membrane distillation (MD) is a thermally driven process in which only vapor molecules are transported through porous hydrophobic membranes. The liquid feed to be treated by MD must be in direct contact with one side of the membrane, but without wetting the membrane to avoid the entrance inside the dry pores [1]. This hydrophobic nature of the membrane prevents the mass transfer in liquid phase and creates a vapor–liquid interface at the pore entrance (Figure 1). In this interface, the volatile compounds in the liquid feed evaporate and diffuse across the membrane pores. On the opposite side of the membrane, the vapor is condensed or removed, depending of the configuration of the MD system [2].

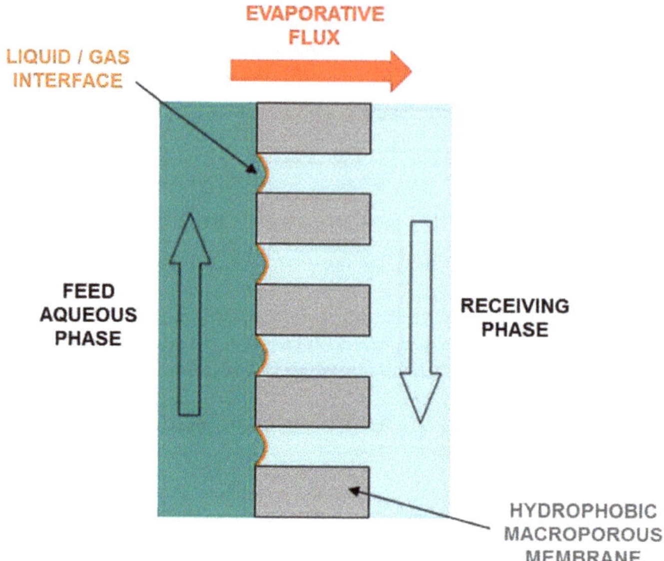

Figure 1. Schematic representation of the performance of membrane distillation.

The benefits of MD compared to distillation or other separation processes based on membranes must be highlighted [3]: the complete theoretical rejection of ions, macromolecules, colloids, cells, and other non-volatiles; lower operating temperatures than conventional distillation; lower operating pressures than conventional pressure-driven membrane separation processes; reduced chemical interaction between membrane and process solutions; less demanding membrane mechanical property requirements; and reduced footprint spaces compared to conventional distillation processes.

The driving force at the origin of the mass transfer through the membrane is a partial pressure gradient, but this gradient can be induced and maintained by different mechanisms, which define the configuration of the MD system. In the case of sweeping gas membrane distillation (SGMD), a cold inert gas sweeps the permeate side of the membrane and removes the vapor molecules, which condensate outside the membrane module (Figure 2). SGMD presents some specific advantages when compared to other MD configurations, like relatively low heat loss by conduction through the membrane, low resistance to mass transfer in the gas phase, and a high driving force for transmembrane transport due to the continuous removal of vapor from the permeate side of the membrane [4].

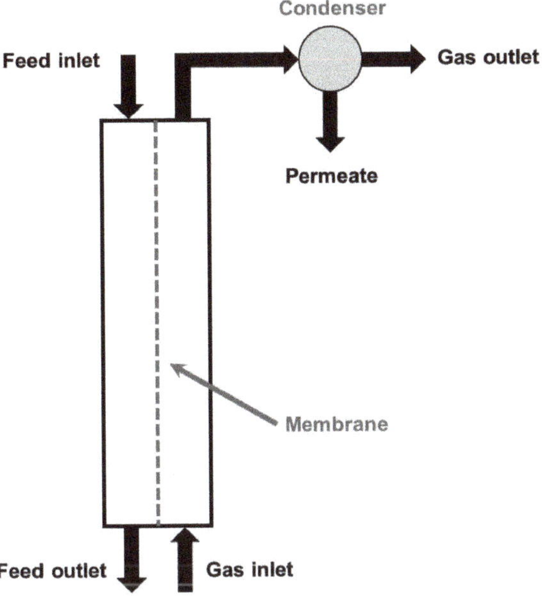

Figure 2. Configuration of a sweeping gas membrane distillation (SGMD) system.

The membranes are key elements for an effective implementation of SGMD systems. Hydrophobic mesoporous–macroporous membranes (pore diameters between 10 nm and 0.5 μm) made of polymeric materials such as polypropylene, polyethylene, polytetrafluoroethylene or polyvinylidenefluoride are generally selected for this application. As previously commented, the hydrophobic character is crucial to avoid the penetration of the aqueous phase through the membrane's porosity. This characteristic can be estimated indirectly from the determination of the intrusion pressure. The general influence of other structural characteristics of the membrane on the performance of SGMD processes is shown in Table 1.

Table 1. Influence of the membrane characteristics on the evaporative flux. + = positive effect; − = negative effect; / = non-referenced effect.

Membrane Characteristics					
Thickness	**Porosity**	**Pore Size**	**Pore Size Distribution**	**Tortuosity**	**Surface Geometry**
− − − −	+ + + +	+ + + +	/	− −	/

Three important parameters of the membrane have direct influence on the evaporative flux: the thickness of the membrane, the porosity, and the size of the pores. First of all, it is clear that the membrane resistance to mass transfer is proportional to the membrane thickness, so the thicker the membrane, the lower the flux. Moreover, it should be noted that the higher the porosity, the higher the flux. Indeed, a membrane having a high porosity offers a greater exchange surface area to the mass transfer and improved diffusivity [5]. Similarly, the evaporative flux increases with the pore diameter of the membrane. However, in order to avoid the penetration of liquid into the membrane, the pore diameter should not be too large. An optimum pore diameter value must therefore be determined for each SGMD application for a good compromise between performance and operation.

The influence of several operating variables on the SGMD processes has been previously investigated [6,7]. The temperatures of the feed and sweeping phases have been identified as the most relevant process conditions; in particular, the temperature difference between the two phases must be taken into consideration. Indeed, SGMD is a thermal process governed by the gradient of partial pressure induced by the difference of temperature on both sides of the membrane. Therefore, the greater the temperature difference between the two phases, the higher the evaporation flux. In addition, the flow velocities of the phases can have an effect on the performance, but it depends on the specific cases [5,8,9]. Table 2 summarizes the influence of the operation conditions on the evaporative flux in the SGMD processes.

Table 2. Influence of the operation conditions on the evaporative flux.

Operation Conditions					
Feed Side			Sweeping Side		
Temperature	Flowrate	Flow Regime	Temperature	Flowrate	Flow Regime
+ + + +	+ +	/	− −	+	/

Desalination has been the most investigated application of MD. This hybrid technology can be employed for the removal of salts and other undesired compounds from a saline water solution to produce freshwater, with quality enough for human consumption, agriculture or industrial uses [10]. Consequently, the development of an improved membrane for this particular application emerges as a very relevant hot topic of research [11,12]. Nevertheless, MD has been successfully implemented in other applications. For example, references of the employment of MD for the dehydration of different solutions, such as aloe vera juice [13], aqueous solutions of glycerol [14] or diethylene glycol [15], can be found in the bibliography. More specifically, SGMD has been applied for the recovery of volatile chemicals like aroma compounds, [16] but examples of its application to the dehydration of glycerol [17] and triethylene glycol [18] have been published as well.

Taking into account this available information, the main aim of this work is the analysis of the potential of SGMD for the concentration of aqueous solutions of 1,3-dimethyl-2-imidazolidinone (DMI). This chemical is a dipolar aprotic solvent with characteristics between tetrahydrofuran (THF) and hexamethylphosphoramide (HMPA), suitable for many types of organic reactions (especially organometallic reactions) [19–27]. Moreover, DMI is used in the manufacture of polymers, as well as in the industry of detergents, dyestuffs, and electronic materials. The recycling of this solvent from aqueous solutions without significant losses is very important for the sustainability of industrial processes. In this work, the concentration from 30% to 50% of aqueous DMI solutions was investigated, paying attention to the main variables of the process: temperatures and flowrates of the feed and sweep gas streams. From these experimental data, simple empirical models were developed to simulate the performance of the installation and calculate the vapor permeation and the DMI losses through the membranes. These empirical models can imply relevant simplification when compared with more complex models developed to represent the performance of SGMD. Finally, the preliminary scaling-up of the process was covered under different multi-objective optimization frameworks.

2. Experimental

2.1. Chemicals

DMI (99.9% of purity) was supplied by Kermel. Ultrapure water (>18.0 MΩ·cm resistivity) was obtained by a Milli-Q Element (Merck KGaA, Darmstadt, Germany).

2.2. Membrane Selection

The selection of a membrane for concentration by SGMD is strictly related to the intended application, as it depends on the nature of the solution to be concentrated. Among the SGMD membranes available on the market, the Liqui-Cel SuperPhobic® membrane (provided by Alting, Hoerdt, France) appears to be the most suitable, since it is made of polypropylene, an inert polymer stable in contact with DMI. The membrane is hydrophobic and intended for applications with solutions characterized by surface tension values between 20 dyne/cm and 40 dyne/cm (in the case of DMI, its value is 40 dyne/cm). The membrane is commercially available as tubular modules of hollow fibers, in several different geometries. The module Liqui-Cel SuperPhobic® X50 was selected and the characteristics of its fibers can be found in Table 3. The total membrane area of the module was 1.2 m^2.

Table 3. Characteristics of the hollow fibers in a Liqui-Cel SuperPhobic® X50 module.

Material	Outer Diameter (µm)	Inner Diameter (µm)	Bubble Point (psi)	Porosity (%)	Pore Diameter (µm)
Polypropylene	300	220	240	40	0.04

2.3. Experimental Installation

On the basis of the characteristics of the module described in the previous section, a SGMD bench installation was designed (Figure 3). Due to the nature of the solution to be concentrated, all of the equipment employed in the bench installation was made of stainless steel in order to be compatible with DMI.

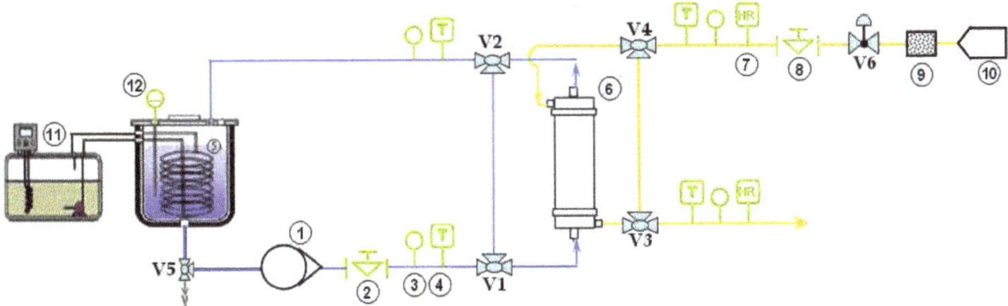

Figure 3. Scheme of the experimental SGMD bench installation.

As shown in Figure 3, the counter current configuration was selected. The DMI solution was stored in a 6 L stainless steel tank (5). During the operation, the volume inside the tank was controlled by a level sensor (12) (Immersion Piezo-Resistive Pressure, Keller 46X). The temperature of the solution was self-regulated by a heat exchanger consisting of a dipping coil (stainless steel), connected to a circulating cryo-thermostat (11). The solution circulation to the module was provided by a gear pump (1) (Micropump M520513), equipped with a speed controller (20–600 L/h). The flow rate was measured by a flow meter (2) with a micro-oval counter (Oval LSN45 LO). The solution passed through the module (6) outside the fibers and was recirculated back to the feed tank after leaving the module. The temperature and pressure were measured at the inlet and outlet of the module by a piezo-resistive pressure transmitter (0–5 bar) and a thermocouple (4), respectively. These same sensors were also installed at the inlet and outlet of the dry air compartment to measure the flow of air circulated in countercurrent inside the fibers. The relative humidity of the air (% RH) was measured at the inlet and outlet of the module by a humidity transmitter (7) (Delta Ohm HD2007) equipped with a remote sensor. Before entering the module, the air from the compressor (10) (7% RH) passed through a desiccant cartridge (9). The airflow was measured by a rotameter (8), in the range of 4–80 L/min (Brooks GT-1000). The airflow was regulated by means of the V6 control valve.

Depending on the position of the valves V1 to V4, the flows of the solution and the air can be done either in closed loop (without passing through the module) or in open loop (passing through the module). This makes it possible to stabilize the operating parameters (temperature, pressure, RH) before starting the concentration step. The V5 valve allows the tank to be drained. Each sensor was connected to a digital display to view the value in real time. The displays were assembled in a box that also provides power to the entire system.

2.4. Experimental Procedures

After the installation and testing of all the capabilities of the bench installation, the characterization of the system performance was carried out by determining its productivity in terms of evaporative flow through the membrane. These initial experiments were carried out by circulating ultrapure water in countercurrent with dry air (1.7% < RH < 5.0%). The resulting productivity was determined under different conditions by varying several parameters (water stream temperature and flowrate, air stream temperature and flowrate), to characterize the most relevant operation conditions. Each time a parameter was varied, the rest of the parameters were kept constant to analyze the influence of each variable individually. The evaporative flow was calculated from the difference in the measured relative humidity in the air stream before and after its pass through the module.

In the case of the concentration of DMI solutions, the experimental procedure was similar to that described for the tests with ultrapure water, with enough stabilization time to attain steady-state conditions. Based on the results of the analysis of the system with water, the operation conditions were chosen to obtain a high yield in terms of evaporative flow and, consequently, a high concentration rate of the DMI solution. The implementation of a valid method for the determination of the DMI concentration was essential for monitoring the performance of the concentration process of the aqueous solution. Moreover, any loss of DMI across the membrane during the concentration process must be identified and quantified. The concentration of DMI solutions was measured by refractometry (the evolution of the refractometer signal as a function of the DMI concentration for a range of 40–3000 ppm was lineal). Therefore, the proposed assay method was validated with a detection limit of 40 ppm.

3. Results and Discussion

3.1. Pure Water Permeation

3.1.1. Determination of the Stabilization Time

The determination of the time required to attain steady-state conditions was the first task to be completed. These experiments were carried out by fixing all parameters of the process (temperature, pressure, water and air flowrates shown in Table 4) and measuring the evolution of the water evaporation flux until stabilization.

Table 4. Conditions of the experiments for the assessment of the stabilization time. RH = relative humidity.

Water Inlet			Air Inlet				Time (h)
Temperature (°C)	**Pressure (bar)**	**Flowrate (L/min)**	**Temperature (°C)**	**Pressure (bar)**	**Flowrate (L/min)**	**RH (%)**	
60	1.05	3	21	0.9	30	1.7	1–5

The water vapor flux measured increases (+5.6%) during the first two hours of operation, reaching 0.09 kg/h·m^2. Then, the evaporation flow after this time until 5 h of operation did not show additional increases, as the values of the evaporative flow obtained were ranged within a margin of error of 3%, which could be attributable to the measurement error of the humidity sensors. According to these results, the stabilization time was estimated to be 2 h. In the following experiments, each measurement point corresponds to the flow value determined after at least a stabilization time of 2 h.

3.1.2. Analysis of the Influence of the Water and Air Temperatures

SGMD is a thermal process governed by a partial pressure gradient as driving force. This gradient is induced by the temperature difference between both compartments separated by the membrane. Therefore, the greater the temperature difference between the two phases, the greater the evaporative flow. This fact justifies the analysis of the influence of the temperature of the water and air streams entering the membrane module.

Since the maximum operating temperature tolerated by the membrane module was 70 °C, the variation of the evaporative flow as a function of the water temperature was investigated in the range of 20 °C to 67 °C (Figure 4). The operation conditions of these tests are equivalent to the ones summarized in Table 4, but with variation of the temperature of the streams.

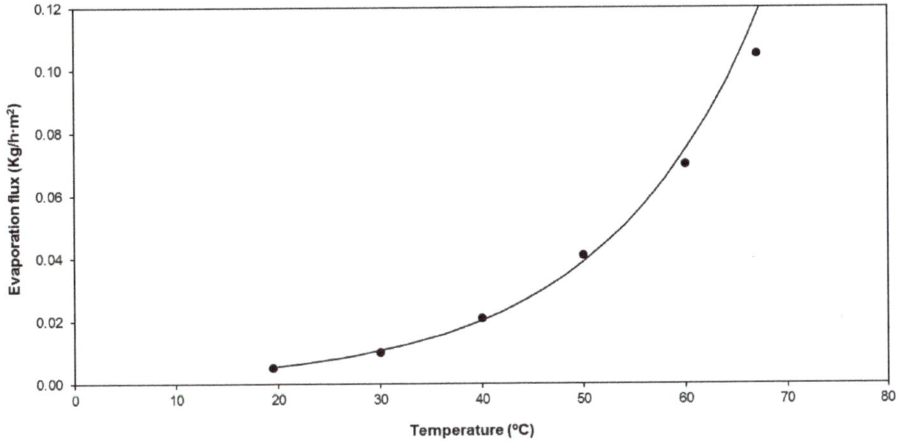

Figure 4. Variation of the evaporative flow as a function of the water temperature.

As shown in Figure 4, the evaporative flow increased with increasing water temperature according to an exponential law under constant air temperature (21 °C). For example, a flux rise of 365% was recorded between 40 °C and 67 °C. This evolution is directly linked to the increase of water vapor pressure as a function of the water temperature which corresponds to the water vapor enhancement described by the Antoine equation [28]. Although the bench installation was not designed to control the temperature of the air stream, the system was prepared to allow the measurement of the evaporative flow at two different inlet air temperatures (21 °C and 26 °C). Under constant water temperature conditions (60 °C), the decrease of air temperature increased the temperature difference between the water and the air, thus increasing the evaporative flow. The variation in air temperature should be considered to be as important as that of the water temperature, since a decrease in the air temperature from 26 °C to 21 °C increased the evaporative flow by 36%.

3.1.3. Analysis of the Influence of the Water and Air Flowrates

According to Khayet [29], the water flowrate at the inlet of the module has practically no effect on the evaporative flow. During the preliminary tests to characterize the bench unit, some experiments were carried out to confirm the insignificant incidence of the water flowrate on the system performance. The obtained results (not shown) confirmed this hypothesis and the same conclusion resulted from experiments with DMI solutions, where the influence of the solution flowrate on the separation process was negligible.

The analysis of the influence of the air flowrate on the evaporative flow revealed it strongly depended on other variables: the temperature of the water appeared as the most relevant operation condition. Therefore, the evaporative flow was determined as a function of the air flowrate at different water temperatures. These operation conditions are summarized in Table 5.

Table 5. Conditions of the experiments for the analysis of the influence of the water and air flowrates.

Water Inlet			Air Inlet			
Temperature (°C)	Pressure (bar)	Flowrate (L/min)	Temperature (°C)	Pressure (bar)	Flowrate (L/min)	RH (%)
20–67	1.05	3	25	0.9	20–34	1.7

The evolution of the evaporative flow as a function of the air flow at different water temperatures is shown in Figure 5. As shown, the evaporative flow was only slightly dependent on the air flowrate, except for the cases where the highest water temperatures were reached. Indeed, in the water temperature range between 20 °C and 50 °C, the evaporative flux was practically constant even if the air flowrate was increased from 20 L/min to 34 L/min.

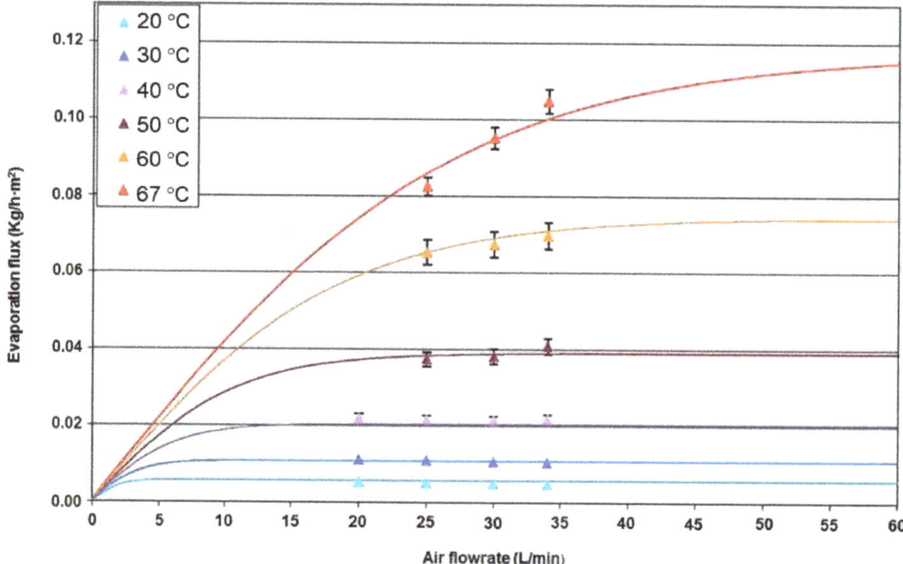

Figure 5. Variation of the evaporative flow as a function of the air flowrate for different water temperatures.

Under the low water temperature conditions (temperatures below 40 °C), an evaporative flow saturation value was reached at 20 L/min. When the feed temperature was increased to 67 °C, the saturation stage is shifted at higher air flow rates. Thus, an increase in evaporative flow is recorded with the air flow rate. The increase is 32% when the air flow rate increases from 25 L/min to 34 L/min (limit of the air flowrate available through the installation).

3.2. Concentration of DMI Solutions

3.2.1. Analysis of the Influence of the Initial DMI Concentration

The effect of the composition of the feed aqueous solution was studied by reporting the evolution of the evaporative flow as a function of the DMI concentration in the feed solution. The operating conditions of these tests are compiled in Table 6.

Table 6. Conditions of the experiments for the analysis of the influence of the initial 1,3-dimethyl-2-imidazolidinone (DMI) concentration (C_{DMI}).

Water Inlet			Air Inlet				C_{DMI} (%)
Temperature (°C)	Pressure (bar)	Flowrate (L/min)	Temperature (°C)	Pressure (bar)	Flowrate (L/min)	RH (%)	
67	1.05	3	25	0.9	34	1.7	0–33

The DMI concentration was varied in the range of 0–33%, and it was shown that the increasing addition of DMI to the feed solution had no effect on the evaporative flow, as the flow maintained a constant value of 0.11 kg/h·m^2.

3.2.2. Analysis of the Losses of DMI by Permeation through the Membrane

During dehydration of an aqueous DMI solution by SGMD, only the water vapor is assumed to pass through the hydrophobic membrane. However, DMI can permeate through the membrane just by steam drag.

In order to detect and quantify any loss of DMI, the air at the outlet of the module was condensed in an immersion trap with liquid nitrogen. The condensate was recovered every 10 min, which made it possible to have a sufficient quantity for DMI quantification. Therefore, at the end of the experiment, the condensed evaporative flux is determined by weighing. This condensed evaporative flux corresponded to a fraction of the total evaporative flux, which was determined by calculation from the registered relative humidity values. Two different sets of experiments were carried out with changes in the DMI compositions, one with 10% and the other with 33%. The total losses of DMI depended on the concentration of the feed solution, since the higher losses corresponded to the more concentrated solution: while loss values of 0.120% were assessed for a 10% DMI feed solution in the reservoir, the losses increased to 0.195% when a 33% DMI feed solution was employed.

3.2.3. Achievement of Concentrated DMI Solutions

The final aim of the experimental tests consisted of the implementation of a concentration test according to a defined specification: attain an aqueous solution with 50% DMI concentration starting from a solution of approximately 30% DMI. The operation conditions for this concentration test are summarized in Table 7.

Table 7. Conditions of the concentration test to obtain a concentrated (50%) DMI solution.

Water Inlet			Air Inlet				Time (h)
Temperature (°C)	Pressure (bar)	Flowrate (L/min)	Temperature (°C)	Pressure (bar)	Flowrate (L/min)	RH (%)	
60	1.05	3	21	0.9	30	1.7	0–9

The evolution of the evaporative flux as a function of the time of the experiment was recorded (not shown). A slight variation of the flux (4.5% reduction from the initial to the final sample) was observed. From this hourly-determined register of the evaporative flux, the evolution of the DMI concentration in the reservoir throughout the experiment was calculated and compared with the samples taken (Figure 6). The target concentration (50%) was reached after 9 h of operation.

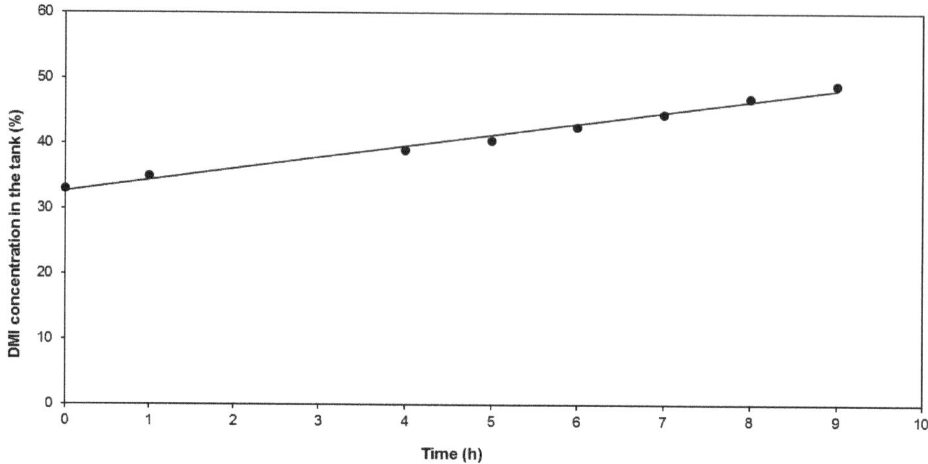

Figure 6. Evolution of the DMI concentration in the feed tank through the time.

The assessment of the amount of DMI that escaped from the module was possible. The DMI loss was expressed as the cumulative mass percentage of DMI in the total condensate mass. Figure 7 shows the evolution of DMI loss as a function of the DMI concentration in the tank, which increased according to an exponential law. It was observed that, at the end of the experimental test (DMI concentration equal to 50%), the cumulative loss was close to 0.55% (this is less than 6 mL per L of evaporated water). This result indicates the possibility to concentrate DMI solutions in spite of the very low vapor permeation flux, which is certainly low for an industrial process if we compare with classical values of membrane distillation processes for water desalination or concentration of salty solutions in literature (from 0.1–10 kg/h·m^2). However, we have to take in consideration that there are not losses of solute in such processes of salty solutions distillation. On the contrary, in the case of the DMI concentration, a critical point concerns DMI losses. These losses are minimized by using a Liqui-Cel SuperPhobic® X50 module which contains hollow fibers with only 40% porosity and 0.04 μm mean pore size. These characteristics are responsible for both the low vapor flux observed and low DMI losses.

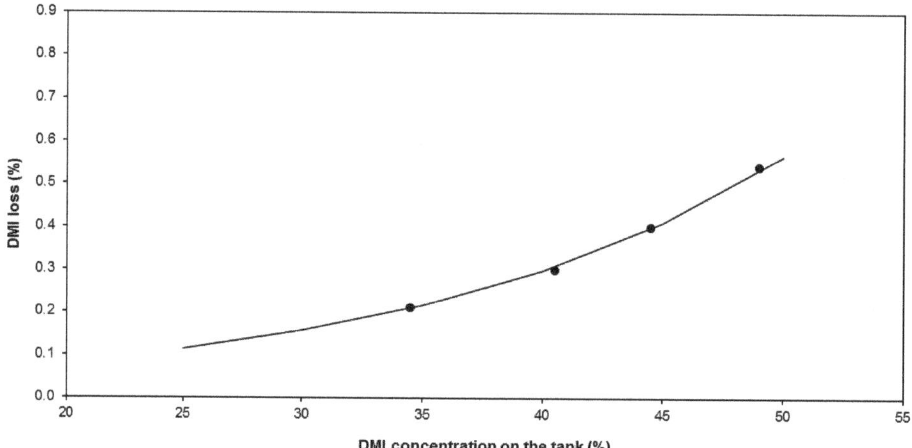

Figure 7. Variation of the DMI losses as a function of the DMI concentration of the feed solution.

3.3. Process Modeling and Scaled-Up Design

Once the technical viability of the process for the concentration of aqueous solutions of DMI by SGMD was demonstrated, the preliminary design of a scaled-up installation was carried out. Although highly robust two-dimensional theoretical models for SGMD simulation have been developed [4,30–32], several authors have preferred the application of empirical models because of their higher simplicity [8,9].

In this work, an empirical model was proposed to describe the performance of the process as a function of the main operation variables: feed inlet temperature T, air flowrate V_{AIR}, and DMI concentration C_{DMI}. The performance of the SGMD system was characterized by the assessment of the fluxes of the two components (water and DMI). On the one hand, the water flux F was calculated with the following equations:

$$F = F_{MAX} \cdot \frac{e^{\frac{V_{AIR}}{V_E}} - e^{-\frac{V_{AIR}}{V_E}}}{e^{\frac{V_{AIR}}{V_E}} + e^{-\frac{V_{AIR}}{V_E}}} \quad (1)$$

$$F_{MAX} = F_0 \cdot e^{\alpha \cdot T} \quad (2)$$

$$V_E = V_0 \cdot e^{\beta \cdot T} \quad (3)$$

where F_{MAX} is the maximal F value that can be attained at one temperature, V_E is the parameter that determines the V_{AIR} influence on the F value (when V_{AIR} equals V_E the F value is $0.762 \cdot F_{MAX}$), F_0 and V_0 and α and β are the corresponding baseline values and exponential parameters of the temperature dependence of the F_{MAX} and V_E values, respectively. On the other hand, the DMI flux F_{DMI} was calculated with the following equations:

$$F_{DMI} = F \cdot \frac{L_{DMI}}{100} \quad (4)$$

$$L_{DMI} = L_0 \cdot e^{\omega \cdot C_{DMI}} \quad (5)$$

where L_{DMI} represents the percentage of DMI losses through the membrane and L_0 and ω represent the baseline value and exponential parameter of the concentration dependence. The values of all the parameters required by the proposed model were obtained after minimization of the errors in the fitting of the experimental data and they are compiled in Table 8. The satisfactory fitting of the experimental data by the proposed model can be observed in Figures 4, 5 and 7.

Table 8. Parameters of the developed SGMD model.

Parameter	Unit	Value
F_0	kg/h·m^2	0.0015
V_0	L/min	0.7265
L_0	%	0.0228
α	°C^{-1}	0.0651
β	°C^{-1}	0.0539
ω		0.0642

Nevertheless, before the preliminary design of a scaled-up installation was done, a complete sensibility analysis of the water flux through the membrane as a function of the feed temperature and the air flowrate was carried out (Figure 8). As the figure shows, the influence of the feed temperature on the performance of the process was more important that the influence of the air flowrate. On the one hand, once certain air flowrate points were attained, values besides these limits resulted in a plateau area where it was not possible to significantly increase the evaporation flux. On the other hand, the exponential influence of the feed temperature resulted in highly improved evaporation fluxes were the maximal considered temperature values were applied. The maximal evaporation flux value was

0.141 kg/h·m^2, which corresponded to 70 °C and 80 L/min, but a value equal to 0.136 kg/h·m^2 was obtained with 70 °C and less than 60 L/min.

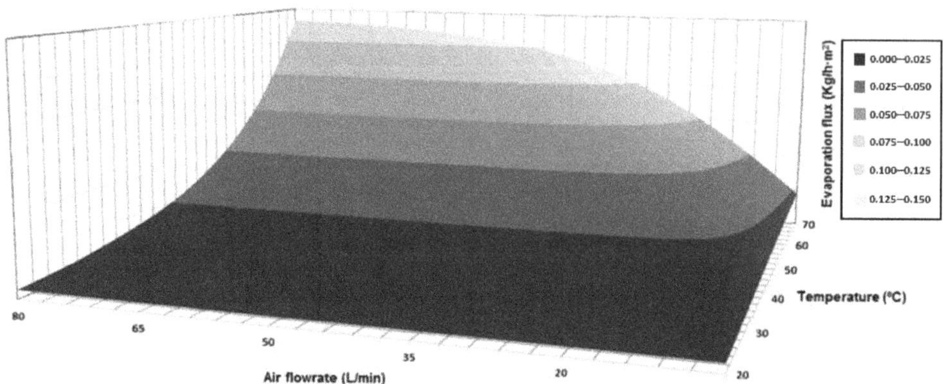

Figure 8. Simulated evaporation flux of the system under different feed temperature and air flowrate conditions.

The preliminary design of a scaled-up installation (based on the use of the membrane modules experimentally characterized in this work) was able to concentrate 3400 kg/h of a 30% DMI solution to achieve a final concentration of 50% DMI. According to the proposed model and the corresponding calculated DMI losses, the installation must be able to evaporate 1368 kg/h of water, which implied 8 kg/h of DMI lost through the membrane. Indeed, 2024 kg/h of a 50% DMI concentrated solution can be obtained.

The optimal design of the installation is not a simple task because of the existence of contradictory objectives. Although the main objective of the design should be the minimization of the total membrane area required, this situation implied maximal feed temperature and air flowrate and, consequently, maximal operation costs. Therefore, 9635 m^2 was the minimal membrane area of the installation when the maximal air flowrate (80 L/min) and feed temperature (70 °C) were applied. The complete evolution of the membrane area required in the installation as a function of the selected air flowrates and feed temperatures can be observed in Figure 9.

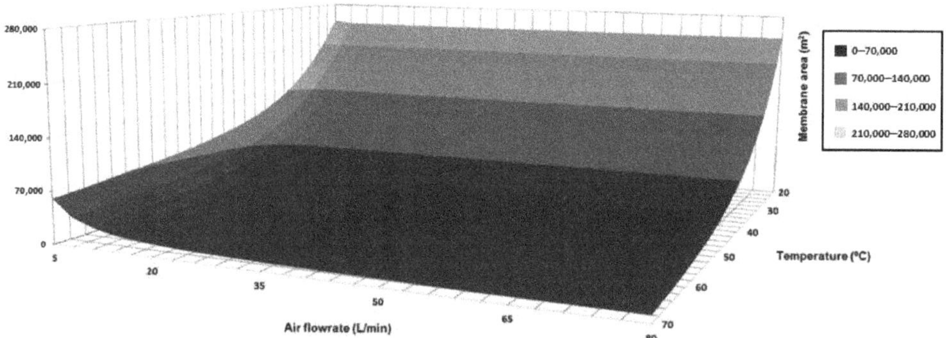

Figure 9. Evolution of the total membrane area required in the installation under different feed temperature and air flowrate conditions.

In order to have a simplified outlook to the multi-objective optimization of the installation, Pareto graphs were prepared. Pareto optimal solutions are solutions that cannot be improved in one objective function without deteriorating the performance in at least another objective [33]. In this case, the epsilon constraint method was employed to obtain Pareto graphs that included the total membrane area compared to total air consumption or total heat requirements [34,35].

Total air consumption was selected as the most adequate variable to take into account the costs due to the selected air flowrate value. The Pareto that resulted from the consideration of the simultaneous minimization of the total membrane and the air consumption is graphed in Figure 10. The results clearly demonstrated that air consumption values above 350 m^3/min only slightly decreased the total membrane area, so air flowrate values below the maximal value analyzed in this work could be selected without excessive detriment in the process performance.

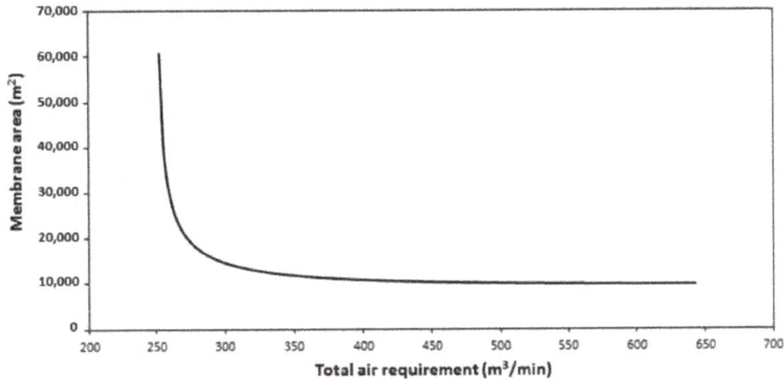

Figure 10. Influence of the total air requirement as sweeping gas on the total membrane area of the process.

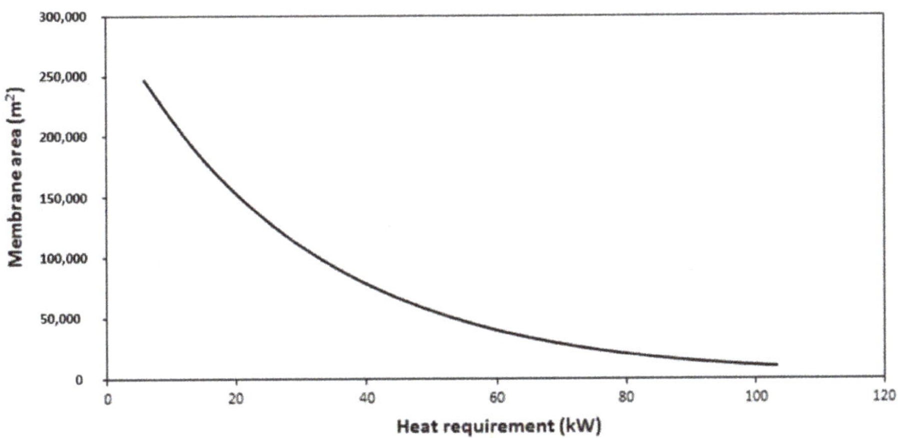

Figure 11. Influence of the heat requirement to increase the feed temperature on the total membrane area of the process.

However, this is not the case for the feed temperature and the corresponding total heat requirement, which was calculated by considering the heat capacity of pure DMI (1.8 J/g·°C) and 60% efficacy in the heat exchanger [36,37]. As shown in Figure 11, the reduction of the heat requirement (by selection of lower feed temperature) implied great membrane area penalties. Therefore, the installation should be designed to work at the maximal feed temperature. Nevertheless, even under these optimal conditions, the resulted total membrane area is excessive. In these circumstances, the scaled-up process cannot be considered competitive and further additional work will be required. These future tasks should consider the identification of more adequate membrane modules, maybe with increased vapor flux, although this should imply increased DMI losses. The design of more complex configurations, with in series stages to recover DMI from the sweep phase, should be taken into account.

4. Conclusions

This work investigated the applicability of sweeping gas membrane distillation (SGMD) to concentrate by dehydration aqueous solutions of 1,3-dimethyl-2-imidazolidinone (DMI). An experimental bench installation equipped with Liqui-Cel SuperPhobic® membranes was employed to analyze the influence of the main operation variables on the process performance. This way, the temperature of the feed stream and the air flowrate were identified as the most relevant variables.

The installation was successfully employed to achieve a concentration of DMI solutions from 30% to 50% under batch conditions. The selected membranes were responsible for the low vapor flux observed but were also effective for the minimization of DMI losses through the membrane since these losses were maintained below 1% of the evaporated water flux. This fact implied that more than 99.2% of the DMI fed to the system was recovered in the produced concentrated solution.

Once the technical viability of the process was confirmed, simple empirical models were developed to simulate the performance of the SGMD process for DMI concentration. These models were applied to the design of a scaled-up installation able to concentrate 3400 kg/h of a 30% DMI solution under continuous operation. The analysis of the influence of the main operation variables was taken into consideration to have a preliminary multi-objective optimization of the system by simultaneous minimization of the total membrane area, the heat requirement and the air consumption. The simulation results show that maximal feed temperature and air flowrate (and the corresponding high operation costs), as well as very high membrane area (9635 m^2) were necessary to reach the contradictory objectives of the process. Under these conditions, the process was deemed to not be competitive for an industrial application and it should be improved before a definitive cost analysis that could be very carefully compared with classical distillation processes.

Author Contributions: R.A., carried out modelling, simulations, wrote and corrected the article; H.S., carried out the experiments; A.D., participated to scientific discussions and corrected the article; C.R., provided the DMI and participated to scientific discussions; J.S.-M., directed the work, participated to scientific discussions and wrote the article.

Funding: This research has not received external funding.

Acknowledgments: R. Abejón acknowledges the financial support from the Spanish Ministry of Economy and Competitiveness (MINECO) through CTQ2014-56820-JIN Project, co-financed by FEDER funds. KERMEL is acknowledged for the financial support of this research.

Conflicts of Interest: The authors declare no conflict of interest.

Nomenclature

C_{DMI}	DMI concentration (%)
F	Evaporated water flux through the membrane (kg/h·m^2)
F_{DMI}	DMI flux through the membrane (kg/h·m^2)
F_{MAX}	Maximal evaporated water flux at certain temperature (kg/h·m^2)
F_0	Baseline evaporated water flux (kg/h·m^2)
L_{DMI}	DMI losses through the membrane (%)
L_0	Baseline DMI losses (%)
T	Feed inlet temperature (°C)
V_{AIR}	Air flowrate (L/min)
V_E	Parameter of the influence of the air flowrate on the evaporated water flux (L/min)
V_0	Baseline value of V_E (L/min)
α	Exponential parameter of the temperature dependence of F_{MAX} (°C^{-1})
β	Exponential parameter of the temperature dependence of V_E (°C^{-1})
ω	Exponential parameter of the concentration dependence of L_{DMI} (°C^{-1})

References

1. El-Bourawi, S.M.; Ding, Z.; Ma, R.; Khayet, M. A framework for better understanding membrane distillation separation process. *J. Membr. Sci.* **2006**, *285*, 4–29. [CrossRef]
2. Curcio, E.; Drioli, E. Membrane distillation and related operations—A review. *Sep. Purif. Rev.* **2005**, *34*, 35–86. [CrossRef]
3. Lawson, K.W.; Lloyd, D.R. Membrane distillation. *J. Membr. Sci.* **1997**, *124*, 1–25. [CrossRef]
4. Karanikola, V.; Corral, A.F.; Jiang, H.; Sáez, A.E.; Ela, W.P.; Arnold, R.G. Effects of membrane structure and operational variables on membrane distillation performance. *J. Membr. Sci.* **2017**, *524*, 87–96. [CrossRef]
5. Karanikola, V.; Corral, A.F.; Jiang, H.; Sáez, A.E.; Ela, W.P.; Arnold, R.G. Sweeping gas membrane distillation: Numerical simulation of mass and heat transfer in a hollow fiber membrane module. *J. Membr. Sci.* **2015**, *483*, 15–24. [CrossRef]
6. Cotamo-de la Espriella, R.J.; Barón-Núñez, F.O.; Muvdi-Nova, C.J. Operating conditions influence on VMD and SGMD for ethanol recovery from aqueous solutions. *CT&F-Ciencia Tecnología y Futuro* **2015**, *6*, 69–80.
7. Shirazi, M.M.A.; Kargari, A.; Tabatabaei, M. Sweeping Gas Membrane Distillation (SGMD) as an alternative for integration of bioethanol processing: Study on a commercial membrane and operating parameters. *Chem. Eng. Commun.* **2015**, *202*, 457–466. [CrossRef]
8. Cojocaru, C.; Khayet, M. Sweeping gas membrane distillation of sucrose aqueous solutions: Response surface modeling and optimization. *Sep. Purif. Technol.* **2011**, *81*, 12–24. [CrossRef]
9. Khayet, M.; Cojocaru, C.; Baroudi, A. Modeling and optimization of sweeping gas membrane distillation. *Desalination* **2012**, *287*, 159–166. [CrossRef]
10. Nakoa, K.; Date, A.; Akbarzadeh, A. A research on water desalination using membrane distillation. *Desalin. Water Treat.* **2015**, *56*, 2618–2630. [CrossRef]
11. Lu, K.J.; Zuo, J.; Chung, T.D. Novel PVDF membranes comprising n-butylamine functionalized graphene oxide for direct contact membrane distillation. *J. Membr. Sci.* **2017**, *539*, 34–42. [CrossRef]
12. Munirasu, S.; Banat, F.; Durrani, A.A.; Haija, M.A. Intrinsically superhydrophobic PVDF membrane by phase inversion for membrane distillation. *Desalination* **2017**, *417*, 77–86. [CrossRef]
13. Dershmukh, S.K.; Sapkal, V.S.; Sapkal, R.S. Dehydration of aloe vera juice by membrane distillation. *J. Chem. Pharm. Sci.* **2013**, *6*, 66–72.
14. Madhumala, M.; Madhavi, D.; Sankarshana, T.; Sridhar, S. Recovery of hydrochloric acid and glycerol from aqueous solutions in chloralkali and chemical process industries by membrane distillation technique. *J. Taiwan Inst. Chem. Eng.* **2014**, *45*, 1249–1259. [CrossRef]
15. Chen, T.H.; Huang, Y.H. Dehydration of diethylene glycol using a vacuum membrane distillation process. *J. Taiwan Inst. Chem. Eng.* **2017**, *74*, 233–237. [CrossRef]
16. Bagger-Jørgensen, R.; Meyer, A.S.; Pinelo, M.; Varming, C.; Jonsson, G. Recovery of volatile fruit juice aroma compounds by membrane technology: Sweeping gas versus vacuum membrane distillation. *Innov. Food Sci. Emerg. Technol.* **2011**, *12*, 388–397. [CrossRef]
17. Shirazi, M.M.A.; Kargari, A.; Tabatabaei, M.; Ismail, A.F.; Matsuura, T. Concentration of glycerol from dilute glycerol wastewater using sweeping gas membrane distillation. *Chem. Eng. Process.* **2014**, *78*, 58–66. [CrossRef]
18. Duyen, P.M.; Jacob, P.; Rattanaoudom, R.; Visvanathan, C. Feasibility of sweeping gas membrane distillation on concentrating triethylene glycol from waste streams. *Chem. Eng. Process.* **2016**, *110*, 225–234. [CrossRef]
19. Leahy, E.M. 1,3-Dimethyl-2-imidazolidinone. In *E-EROS Encyclopedia of Reagents for Organic Synthesis*; Paquette, L.A., Ed.; Wiley: New York, NY, USA, 2001.
20. Seki, T.; Kokubo, Y.; Ichikawa, S.; Suzuki, T.; Kayaki, Y.; Ikariya, T. Mesoporous silica-catalysed continuous chemical fixation of CO_2 with *N,N′*-dimethylethylenediamine in supercritical CO_2: The efficient synthesis of 1,3-dimethyl-2-imidazolidinone. *Chem. Commun.* **2009**, 349–351. [CrossRef]
21. Zhao, L.M.; Zhang, S.Q.; Jin, H.S.; Wan, L.J.; Dou, F. Zinc-mediated highly α-regioselective prenylation of imines with prenyl bromide. *Org. Lett.* **2012**, *14*, 886–889. [CrossRef]
22. Kitamura, T.; Gondo, K.; Katagiri, T. Synthesis of 1,2-bis(trimethylsilyl)benzene derivatives from 1,2-dichlorobenzenes using a hybrid metal Mg/CuCl in the presence of LiCl in 1,3-dimethyl-2-imidazolidinone. *J. Org. Chem.* **2013**, *78*, 3421–3424. [CrossRef] [PubMed]

23. Li, Y.F.; Yuan, Y.P.; Wang, K.K.; Jia, J.; Qin, X.L.; Xu, Y. Conversion of sawdust into 5-hydroxymethylfurfural by using 1,3-dimethyl-2-imidazolidinone as the solvent, Iran. *J. Chem. Chem. Eng.* **2013**, *3*, 75–79.
24. Hou, Q.; Li, W.; Ju, M.; Liu, L.; Chen, Y.; Yang, Q.; Wang, J. Separation of polysaccharides from rice husk and wheat bran using solvent system consisting of BMIMOAc and DMI. *Carbohydr. Polym.* **2015**, *133*, 517–523. [CrossRef]
25. Qian, Q.; Cui, M.; Zhang, J.; Xiang, J.; Song, J.; Yang, G.; Han, B. Synthesis of ethanol via reaction of dimethyl ether with CO_2 and H_2. *Green Chem.* **2018**, *20*, 206–213. [CrossRef]
26. Tian, M.; Wang, J.; Zhou, J.; Han, J.; Du, F.; Xue, Z. Hydrogen bonding improved atom transfer radical polymerization of methyl methacrylate with a glycerol/1,3-dimethyl-2-imidazolidinone green system. *J. Polym. Sci. A Polym. Chem.* **2018**, *56*, 282–289. [CrossRef]
27. Xie, L.; Cho, A.N.; Park, N.G.; Kim, K. Efficient and reproducible $CH_3NH_3PbI_3$ perovskite layer prepared using a binary solvent containing a cyclic urea additive. *ACS Appl. Mater. Interfaces* **2018**, *10*, 9390–9397. [CrossRef]
28. Hengl, N.; Mourgues, A.; Belleville, M.P.; Paolucci-Jeanjean, D.; Sanchez, J. Membrane contactor with hydrophobic metallic membranes: 2. Study of operating parameters in membrane evaporation. *J. Membr. Sci.* **2010**, *355*, 126–132. [CrossRef]
29. Khayet, M.; Godino, M.P.; Mengual, J.I. Thermal boundary layers in sweeping gas membrane distillation processes. *AIChE J.* **2002**, *48*, 1488–1497. [CrossRef]
30. Charfi, K.; Khayet, M.; Safi, M.J. Numerical simulation and experimental studies on heat and mass transfer using sweeping gas membrane distillation. *Desalination* **2010**, *259*, 84–96. [CrossRef]
31. Plaza, A.; Merlet, G.; Hasanoglu, A.; Isaacs, M.; Sanchez, J.; Romero, J. Separation of butanol from ABE mixtures by sweep gas pervaporation using a supported gelled ionic liquid membrane: Analysis of transport phenomena and selectivity. *J. Membr. Sci.* **2013**, *444*, 201–212. [CrossRef]
32. Perfilov, V.; Fila, V.; Sanchez-Marcano, J. A general predictive model for sweeping gas membrane distillation. *Desalination* **2018**, *443*, 285–306. [CrossRef]
33. Abejón, R.; Belleville, M.P.; Sanchez-Marcano, J. Design, economic evaluation and optimization of enzymatic membrane reactors for antibiotics degradation in wastewaters. *Sep. Pur. Technol.* **2015**, *156*, 183–199. [CrossRef]
34. Abejón, R.; Garea, A.; Irabien, A. Multiobjective optimization of membrane processes for chemicals ultrapurification. *Comput. Aided Chem. Eng.* **2012**, *30*, 542–546.
35. Mavrotas, G. Effective implementation of the e-constraint method in multiobjective mathematical programming problems. *Appl. Math. Comput.* **2009**, *213*, 455–465.
36. Fakheri, A. The shell and tube heat exchanger efficiency and its relation to effectiveness. In Proceedings of the ASME 2003 International Mechanical Engineering Congress and Exposition, Washington, DC, USA, 15–21 November 2003; pp. 9–15.
37. Sadeghzadeh, H.; Aliehyaei, M.; Rosen, M.A. Optimization of a finned shell and tube heat exchanger using a multi-objective optimization genetic algorithm. *Sustainability* **2015**, *7*, 11679–11695. [CrossRef]

Article

Evaluation of Permeate Quality in Pilot Scale Membrane Distillation Systems

Alba Ruiz-Aguirre [1,2,*], Juan A. Andrés-Mañas [2] and Guillermo Zaragoza [3,*]

1 Dipartimento di Ingegneria, Università degli Studi di Palermo (UNIPA), viale delle Scienze, Ed. 6, 90128 Palermo, Italy

2 Department of Chemical Engineering, Universidad de Almeria—CIESOL, 04120 Almería, Spain; juanantonio.andres@psa.es

3 CIEMAT—Plataforma Solar de Almeria, Ctra. de Senés s/n, 04200 Tabernas, Almería, Spain

* Correspondence: alba.ruiz@psa.es (A.R.-A.); guillermo.zaragoza@psa.es (G.Z.)

Received: 29 April 2019; Accepted: 3 June 2019; Published: 5 June 2019

Abstract: In this work, the salinity of permeate obtained with membrane distillation (MD) in pilot scale systems was analyzed. Experiments were performed with three different spiral-wound commercial modules, one from Solar Spring with 10 m^2 surface membrane area and two from Aquastill with 7.2 and 24 m^2. Intermittent operation meant that high permeate conductivity was measured in the beginning of each experiment, which was gradually decreasing until reaching a constant value (3–143 $\mu S \cdot cm^{-1}$ for seawater feed). The final quality reached did not depend on operating conditions, only the time it took to reach it. This can be because the permeate flux dilutes the minimal feed leak taking place through pinholes in the membranes. Larger feed leak through the membrane was observed when operating in vacuum-enhanced air-gap MD configuration (V-AGMD), which is compatible with this explanation. However, for the increase of feed leak with salinity (up to 1.8 M), a conclusive explanation cannot be given. Pore wetting due to crystallization is discarded because the high permeate quality was recovered after washing with distilled water. More studies at higher salinities and also at membrane level are required to investigate this.

Keywords: desalination; brine treatment; membrane distillation; pilot scale; permeate quality; membrane filtration; high salinity

1. Introduction

Seawater desalination has become the main option used to alleviate the problem of water scarcity in coastal regions [1]. Out of all the different desalination processes, membrane technologies are the most widely used throughout the world. Among them, reverse osmosis (RO) is the most studied and industrially implemented method because of its economic competitiveness compared to other processes [2–4]. However, the problems related to the discharge of concentrated brines into the environment have not yet been resolved and are gaining increasing relevance [5]. In recent years, science and technology of membranes have improved in trying to find solutions to these environmental problems in order to achieve the goal of zero liquid discharge (ZLD) [6,7]. Additionally, this concept offers the opportunity of recovering valuable materials from desalination brines after a process of membrane crystallization [8–12]. Since pressure-driven membrane processes such as RO cannot treat concentrated brines because of their high osmotic pressures, thermal desalination technologies must be used [13,14]. A non-isothermal membrane process such as membrane distillation (MD) appears as a suitable solution. MD has been proposed to treat aqueous solutions of high salt concentration derived from other desalination processes, for producing more water improving the quality of the RO product water, and for reducing the volume of waste in order to achieve ZLD [8]. The main challenge of MD technology in this context is the possibility of irreversible pore wetting of the membrane resulting in a considerable decline of MD performance and ultimately the subsequent discard of the membrane. However, not many studies have been performed with MD for treating high salinity solutions, and the majority of them have been carried out at a laboratory scale. Experimental works with high concentration aqueous solutions of NaCl have been carried out using air-gap MD [15,16], direct contact MD [17,18] and vacuum MD [19], but they have all been done at a very small scale, typically for surface membrane areas of 0.014 m^2 or lower. Also for the treatment of RO brines with MD, experiments have been mostly performed at lab scale, using DCMD [20,21] and VMD [22], with membrane areas of 0.014, 0.04 and 0.16 m^2, respectively. Even when coupling membrane distillation with crystallization [11,22–24], the scale of the experiments is usually too small for extrapolating to commercial modules. As many authors state, experiments at pilot scale are needed [25]. Few studies are reported, Minier-Matar et al., [26] compared two commercial MD technologies at pilot scale (membrane areas of 6.4 and 4.6 m^2, respectively) for treating brines from thermal desalination (70 g/L total dissolve solids), obtaining a high quality distillate (< 10 $\mu S \cdot cm^{-1}$) with the first one (vacuum-enhanced 4-effect MD configuration). Duong et al [27] studied the behaviour of a pilot MD system from Aquastill (membrane area 7.2 m^2) for the treatment of RO brine from coal seam gas (CSG) produced water. The permeate conductivity was reported to be 500 $\mu S \cdot cm^{-1}$, which is higher than that obtained in lab-scale studies. Winter et al [28] characterized the performance of the Oryx 150 module from Solar Spring (membrane area 10 m^2) for different feed salinities (0–1.8 M) without focusing on the quality of the permeate. However, they reported larger values than in lab-scale experiments, as also done by Ruiz-Aguirre et al., [29] using a similar module but with simulated seawater (0.6 M). Finally, Schwantes et al [30] studied an AGMD module adapted with an active draining of the air gap by low pressure air blowing using a range of concentration on the feed between 0 $g \cdot kg^{-1}$ and 240 $g \cdot kg^{-1}$. Although the study was focused on the characterization of flux, GOR and thermal efficiency, permeate conductivity values were shown. Comparing results without and with blower, the drainage of the air gap with pressurized air, improved the overall average of the quality of permeate and the increase of permeate conductivity with higher feed salinity was significantly mitigated. There are not many other studies of MD at pilot-scale where the permeate quality has been analyzed in depth.

In this work, a Solar Spring module and two different Aquastill modules were used to study the quality of permeate considering different feed concentrations in a range including high salinity, to test whether these commercial modules are suitable for industrial implementation. These modules have been fully characterized in terms of permeate production and energy efficiency [30–32], but extensive information on permeate quality has not been provided yet.

2. Materials and Methods

In this study, three different MD commercial spiral wound modules were evaluated: Oryx 150, AS7, and AS24. The Oryx 150 module was designed by Solar Spring (Freiburg, Germany) in collaboration with Fraunhofer Institute for Solar Energy Systems ISE (Fraunhofer ISE, Freiburg, Germany), while the AS7 and the AS24 modules were built by Aquastill company (Sittard, Netherlands). The former was operated in permeate-gap (PGMD) configuration, while the latter two were operated both in air-gap (AGMD) and vacuum-enhanced air-gap (V-AGMD) configurations. Table 1 summarizes the main characteristics of the three modules.

To carry out the experiments, synthetic solutions were prepared as feed, using marine salts obtained from the natural saltworks of Cabo de Gata (Almería, Spain). Feed solution for the different concentration tested (in a range from 0.6 M to 2.4 M) was prepared dissolving these salts in demineralized water. Demineralized water was produced from brackish water using RO and electrodeionization. The chemical properties of demineralized water were: conductivity: 2 $\mu S \cdot cm^{-1}$, pH: 6.3 and turbidity 0.1 NTU. The dissolved total carbon (DTC) was 1.33 $mg \cdot L^{-1}$ and dissolved inorganic carbon (DIC) was 0.42 $mg \cdot L^{-1}$.

The operation of the commercial MD modules consisted of pumping firstly the feed water from the feed tank into the cold channel (see Figure 1). The feed, pre-heated with the latent heat of condensation, was then additionally heated up in a heat exchanger connected to a solar thermal field, and then it flowed along the hot channel, ending up in the feed tank for recirculation. Permeate was also returned to the feed tank to keep a constant salinity, but three samples were collected every 15 min for measuring its conductivity. Since residual heat of both currents overheats the feed tank, the feed water was cooled with a compressor chiller before entering the module again. More details are given elsewhere for the specific operation of the SolarSpring [33] and the Aquastill [32] modules in this pilot plant.

Table 1. Characteristics of the membrane distillation (MD) commercial modules used in this study.

Characteristics	Oryx 150	AS7/AS24
Configuration:	PGMD	AGMD and V-AGMD
Channels length:	7 m	1.5 m/5 m
Channels thickness:	3.2 mm	2 mm.
Membrane surface area:	10 m^2	7.2 m^2/24 m^2
Membrane material:	active layer: PTFE; support: PP	active layer: LDPE [1]
Active layer thickness:	70 μm	76 μm
Active layer pore size:	0.2 μm	0.3 μm
Active layer porosity:	80%	85%
Support layer thickness:	280 μm	-
Support layer porosity:	50%	-
Gap thickness:	1 mm	0.88 mm
Vacuum pressure:	-	150–250 mbar abs

[1] no support.

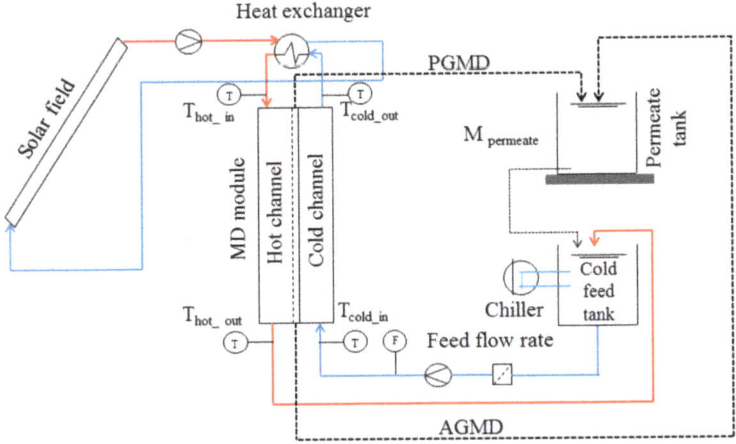

Figure 1. Process flow diagram of the membrane distillation (MD) operation.

In the Oryx 150 module, all the inlets and outlets were located on the top of the module, so the module was left filled with the feed between the tests. In the AS7 and the AS24 modules, the cold inlet channel and the hot outlet channel were placed at the bottom of the module. Therefore, to avoid the emptying of the module when tests were not performed, the circulation of the feed through the module was not interrupted, only reduced and continued without the application of heat.

In this study, a series of experiments with a different hot inlet temperature (T_{hot}), cold inlet temperature (T_{cold}), feed flow rate (F), and feed salinity (S) were considered. In total, a large number of experiments were performed along a large period of time (total operation time of the pilots exceeded 300 h). Table 2 summarizes the combination of the operating conditions in the experiments carried out with the different modules. The flow rate setpoints were achieved by using a variable frequency drive connected to the circulation pump. The changes were done gradually, and the hydraulic pressure inside the module was monitored continuously to prevent it from exceeding the maximum value given by the module manufacturers (700 mbar in the case of SolarSpring and 600 mbar in the case of Aquastill), with the control system acting accordingly. Thus, the hydraulic pressure on the surface membrane was kept more than five times below the nominal liquid entry pressure of the membranes (4 bar in the case of SolarSpring, 3.8 bar in the case of Aquastill). In the AS7 and the AS24 modules, experiments were carried out in standard AGMD and V-AGMD configuration. In the latter case, the suction of air from the gap resulted in absolute pressure in the gap channel varying between 150 mbar and 250 mbar. Experiments with each combination of operating conditions were carried out three times to guarantee the statistical validity of the results.

Table 2. Operating conditions tested on the three commercial MD modules.

Variable (Unit)	Module	Range of Operation
T_{evap} (°C)	All modules	60–80
T_{cond} (°C)	All modules	20–30
F ($L{\cdot}h^{-1}$) *	Oryx 150	400–600
	AS7, AS24	400–1100
S (M) **	Oryx 150	0.6–1; 1–1.5; 1.5–2.4
	AS7	0.6–1.2; 1.2–1.8; 1.8–2.4
	AS24	0.6–1.2; 1.2–1.8

* The highest limit of F was established by the maximum hydraulic pressure allowed. ** S range was chosen considering recommendations of the module manufacturers.

The permeate quality was analyzed in terms of the conductivity, measured with conductivity meter Portavo 902 (version COND, Knick), conveniently calibrated for the appropriate salinity ranges. In order to remove the influence of temperature in these measurements and facilitate their comparison, temperature reference was set to 20 °C. Samples of around 40 mL each 15 min were taken and measured twice. In the case of the Aquastill V-AGMD system, sampling time cannot be fixed, because permeate was discharged discontinuously by the system with a frequency that was very dependent on operational conditions. The conductivity meter provides directly the value of electrical conductivity of the solution referred to 20 °C. Finally, conversion from electrical conductivity (EC) in $mS{\cdot}cm^{-1}$ to concentration (c) in $g{\cdot}L^{-1}$ was made (see Appendix A).

Salt Rejection Factor (SRF) was calculated with the following expression:

$$SRF(\%) = \frac{c_f - c_p}{c_f} \cdot 100, \quad (1)$$

where c_f is the feed concentration in $g{\cdot}L^{-1}$ and c_p is the permeate concentration also in $g{\cdot}L^{-1}$.

The membrane leak ratio, representing the ratio of the feed that passes through the membrane, was calculated following Equation (2). A normalized membrane leak ratio (to membrane surface area) was also calculated (see Equation (3)).

$$\text{Membrane leak ratio } (\%) = \frac{\text{Permeate flow rate } \left(L{\cdot}h^{-1}\right)}{F\left(L{\cdot}h^{-1}\right)} \cdot \frac{C_p\left(g{\cdot}L^{-1}\right)}{C_f\left(g{\cdot}L^{-1}\right)} \cdot 100, \quad (2)$$

$$\text{Normalized membrane leak ratio } \left(\%{\cdot}m^{-2}\right) = \frac{\text{Membrane leak ratio } (\%)}{\text{Surface Membrane Area } (m^2)}, \quad (3)$$

3. Results and Discussion

3.1. Evolution of the Quality of Permeate along the Operation with the Oryx 150 Module

Figure 2 shows the evolution of the conductivity (left) and SRF (right) of permeate obtained during two different days of operation with the Oryx 150 module for a feed salinity of 0.6 M. The curve represented with blue squares illustrates one day when it was operated with two different operating conditions. During the first 25 min, the variables were adjusted to the desired ones (F = 500 L·h^{-1}, T_{hot} = 70 °C and T_{cold} = 30 °C). Later, it operated during 75 min in those conditions. After that, another experiment started increasing T_{hot} by 5 °C with respect to the previous experiment. These conditions were also kept during 75 min after a transition period of 25 min. The conductivity of the first permeate sample obtained was 40,300 μS·cm^{-1} (equivalent to a salt concentration of 27 g·kg^{-1}), that would correspond to a SRF of 23.4%. However, after only 5 min, the conductivity decreased considerably (600%) to reach 6000 μS·cm^{-1} (3 g·kg^{-1}). After 50 min from the first permeate, the conductivity was 239 μS·cm^{-1}, already below the taste threshold established by the World Health Organization (WHO): for sodium chloride in the range 0.2–0.3 g·kg^{-1}, which corresponds to a conductivity of 400–600 μS·cm^{-1} [34]. This WHO standard was taken only as a reference, not implying that permeate produced in MD can be drinkable (as a matter of fact, for that purpose the permeate would generally need remineralization). The corresponding SRF for this value was higher than 99%. The conductivity continued to decrease, reaching a final value of 5 μS·cm^{-1} (0.003 g·kg^{-1}) corresponding to a SRF value close to 100% (99.99%). The curve represented with red circles corresponds to another day when the module was operated with two different conditions. As in the previous case, the first 25 min were used to adjust the operating variables. After that and during 75 min, the module was operated at F = 400 L·h^{-1}, T_{hot} = 60 °C and T_{cold} = 20 °C. Later, the T_{hot} was changed to 65 °C keeping the rest of the variables during 75 min constant. The conductivity of the first sample was 33,750 μS·cm^{-1} (21.7 g·kg^{-1}). As in the previous case, the quality of the first permeate was low (equivalent SRF would be 35%). The conductivity decreased to 1515 μS·cm^{-1} (1 g·kg^{-1}) 45 min after obtaining the first permeate. After 45 min, the conductivity reached a value of 350 μS·cm^{-1} (0.18 g·kg^{-1}). In this day, a very low value of conductivity was also reached finally (4 μS·cm^{-1} (0.002 g·kg^{-1})). The trend of the two curves was the same, namely, a high conductivity in the first permeate and then a decrease to lower values throughout the daily operation. The high conductivity when starting the operation is often explained by the formation of crystals inside the pores, when the operation is stopped [35]. These crystals would be washed as permeate is produced when restarting the operation, therefore the improvement of the permeate conductivity along the operation. However, crystallization inside the pores would cause permanent damage, as is the case for crystallization fouling [36,37]. The fact that excellent permeate quality is obtained consistently after several months of operation when this initial bad permeate quality takes place every day, makes this explanation unlikely. Another possibility is the leaking of feed solution through possible pinholes existing in the membrane. When the operation ends, the flux of vapor is interrupted and only the small passage of liquid feed through the defects of the membrane is collected in the permeate channel. When the operation restarts, permeate has a high salinity because it contains the feed liquid. Since the fraction of feed liquid that passes through the membrane is low (membrane leak ratio of 10^{-5}%), as more permeate is produced this leak is diluted in the condensation channel and the permeate conductivity decreases.

Similar results were reported in other studies [28,35]. However, in the study carried out by Winter et al. [19] with a similar module, the conductivity level decreased faster at the beginning of the experiment than in this work. This could be due to different operating conditions. As illustrated in Figure 2, the final permeate quality did not depend on the operating conditions, as expected from the results obtained in other studies [38]. The influence of the operating conditions was on the time necessary to reach conductivity below the taste threshold. When operating conditions drove to higher permeate production, the time needed was lower. For example, in the first day, the necessary time was 50 min; in the second day, 89 min were needed. This was due to the fact that in the first day the permeate production was greater than in the second one (Table 3). The reduction of the permeate production (40%) from one day to another was almost proportional to the increase of the time (44%) necessary to achieve a good permeate quality (WHO taste threshold). The objective of this work was not to analyze the permeate flux in the modules, which was already studied in other publications [31,32]. However, it is worth stressing that the low values shown in Table 3 are a common feature in all commercial modules, and is explained because of the lower driving force as a consequence of the higher heat transfer through the membrane along the longer channels compared to lab-scale modules. Fluid velocity in the feed channel of the Oryx 150 module was between 0.08 $m \cdot s^{-1}$ and 0.11 $m \cdot s^{-1}$ (corresponding Reynolds numbers 155 and 230, respectively) [31].

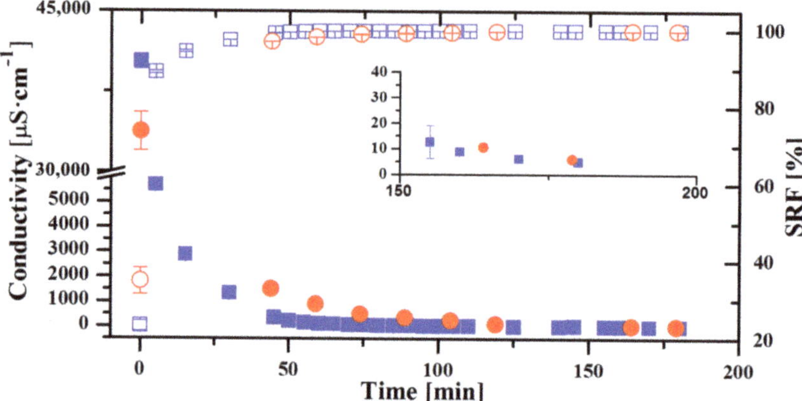

Figure 2. Permeate conductivity and Salt Rejection Factor (SRF) obtained in two typical days of operation with the module Oryx 150. Blue squares: conductivity of day 1; red circles: conductivity of day 2; blue empty squares: SRF of day 1; red empty squares: SRF of day 2.

Table 3. Permeate production with the operating conditions of the cases shown in Figure 2.

Day	F ($L \cdot h^{-1}$)	T_{hot} (°C)	T_{cold} (°C)	P_{flux} ($L \cdot h^{-1} \cdot m^{-2}$)
1	500	70	30	1.5
2	400	60	20	0.9

Figure 3 shows the permeate conductivity obtained in an experiment performed after the module had been left filled with demineralized water subsequently to a cleaning with this water at 80 °C. The experiment was carried out with F = 400 $L \cdot h^{-1}$, T_{hot} = 80 °C and T_{cold} = 20 °C. Permeate conductivity was lower than 2 $\mu S \cdot cm^{-1}$ (0.001 $g \cdot kg^{-1}$) since the beginning and throughout all the experiment, being SRF practically 100%. Therefore, cleaning and leaving the module filled with demineralized water was enough to avoid the high conductivity at the beginning of the experiment. The experiment started shortly after filling the module with the feed solution. Although leak of the feed through the pinholes may have occurred, this was not accumulated before permeate production took place from the vapor flux and so, the conductivity spike was not observed. This helps validate the hypothesis that the high salinity at the beginning of the experiment was due to the leak of liquid feed through membrane defects as pinholes.

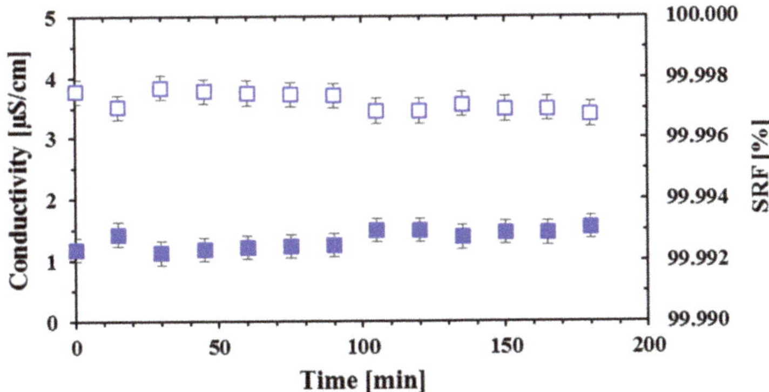

Figure 3. Permeate conductivity and SRF obtained with the module Oryx 150 after cleaning with demineralized water at 80 °C. Blue squares: conductivity, blue empty squares: SRF.

3.2. Evolution of the Permeate Quality along the Operation with Aquastill Modules

Figure 4 shows the evolution of the conductivity (left) and SRF (right) for a typical day of operation with the AS7 (squares) and the AS24 (circles) modules. These two days are representative of the permeate quality obtained throughout the operation with these modules for a salinity of 0.6 M. In both curves, the initial time corresponds to the moment when the first permeate was collected. In the AS7 module, the curve corresponded to one day in which two different operating conditions were tested with their corresponding stabilization time. Specifically, it was first operated with F = 400 $L{\cdot}h^{-1}$, T_{hot} = 70 °C and T_{cold} = 20 °C. After that, F was increased up to 500 $L{\cdot}h^{-1}$. Regarding the AS24 module, operating conditions tested the reference day were firstly F = 500 $L{\cdot}h^{-1}$, T_{hot} = 80 °C and T_{cold} = 20 °C, later F was increased to 600 $L{\cdot}h^{-1}$, and finally T_{hot} was varied to 70 °C. At the beginning, as in the Oryx 150 module, the conductivity was very high; though later the values were lower than the taste threshold established by WHO. In the AS7 module, after 50 min the permeate conductivity was 46 $\mu S{\cdot}cm^{-1}$ and in the AS24 the taste threshold was reached after 40 min. The conductivity of the samples continued to decrease, though the minimum value reached with the AS24 module was greater than that of the Oryx 150 and the AS7 modules. While in the AS7 module the conductivity dropped to a value of 3 $\mu S{\cdot}cm^{-1}$ (SRF practically 100%), the minimum permeate conductivity in the AS24 module was 140 $\mu S{\cdot}cm^{-1}$. The corresponding SRF was 99.7%. Although this value was lower than that obtained in the Oryx 150 and the AS7, it is still higher than that obtained with other desalination technologies (between 95% and 98% for SWRO [39]). As in the case of the Oryx 150 module, the final permeate quality did not depend on the operating conditions. The worse final permeate quality obtained in the AS24 could be due to the greater membrane surface and therefore higher probability of having defects on the membrane as pinholes. The membrane leak ratio was about 10^{-4}% for the AS7 module and 10^{-3}% for the AS24 module. Moreover, the AS7 module had larger permeate flux than the AS24 module (Table 4) [32]. The worse permeate quality in the AS24 module limits its use in certain applications. For example, the conductivity of permeate required in the refrigeration circuits of diesel engines or in the preparation of urea to reduce NO_x emissions of a thermal power plant has to be lower than 100 $\mu S{\cdot}cm^{-1}$ [40].

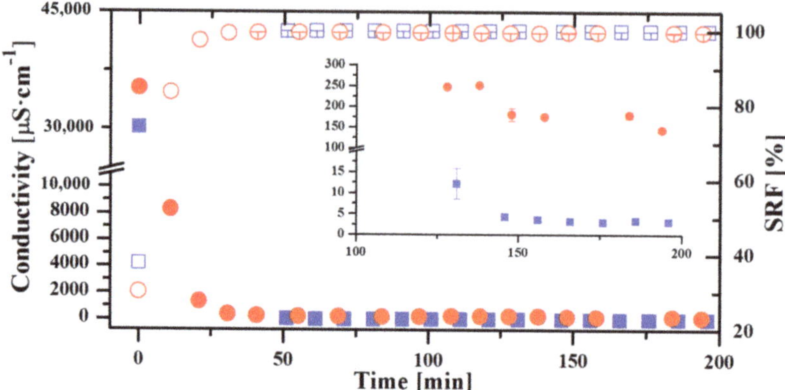

Figure 4. Permeate conductivity and SRF in an operation day with the AS7 module (blue squares and blue empty squares respectively) and the AS24 module (red circles and red empty circles respectively) both in AGMD configuration.

Table 4. Permeate production with the operating conditions of the cases shown in Figure 4.

Module	F (L· h^{-1})	T_{hot} (°C)	T_{cold} (°C)	P_{flux} (L·h^{-1}·m^{-2})
AS7	500	70	20	2.7
AS24	600	80	20	1.1
AS24	600	80	20	1.3
AS24	600	70	20	1.1

3.3. Effect of the Feed Salinity on the Permeate Quality

The previous tests were done for feed salinity equivalent to seawater. Once established that the operating conditions (temperatures and flow rate) do not affect the final permeate quality obtained with each module, the focus of the work was put on the analysis of the influence of the feed salinity on the different modules. Figure 5 shows the SRF for the three MD commercial modules analyzed in this study considering both configurations (AGMD and V-AGMD) in the AS modules for different initial feed concentrations. Operating conditions were nominal (optimized for maximum permeate production) for all feed concentrations. For the AS7 and the AS24 modules T_{hot}, T_{cold} and F were 80 °C, 25 °C and 1100 L·h^{-1} respectively, and for the Oryx 150 module, the only change was that F was 600 L·h^{-1}. The Oryx 150 module showed excellent separation performance with a SRF always above 99.99% for the different feed concentrations tested. The two Aquastill modules tested produced a good quality of permeate for a feed concentration of 0.6 M in both AGMD and V-AGMD configurations. For feed concentration of 1.2 M, only the AS7 module in AGMD configuration showed a SRF according to the WHO standard value. For the rest of feed concentrations, AS modules showed a SRF lower than that, and worsening as the feed concentration increased. When both configurations (AGMD and V-AGMD) were compared, the application of vacuum in the gap affected negatively the permeate quality. Moreover, the effect of the vacuum was slightly more pronounced in the AS24 module than in the AS7. For example, for feed concentration 1.8 M, while in the AS7 module the SRF reduction reached 0.6%, in the AS24 module it amounted to 1%. The application of vacuum reduces the mass transfer resistance in the gap, favouring the passage of liquid feed through the pinholes. This effect was slightly stronger in the AS24 module, possibly due to its larger surface membrane area. Also, the permeate quality decreased as the salinity of the feed (and therefore that of the leak) increased. This effect was also shown by Schwantes et al. [30]. To better assess this matter, the membrane leak ratio was calculated and compared for the different modules (Figure 6). The maximum value was around 0.12% in the AS24 module when operating in V-AGMD configuration for the highest salinity.

For this module, reliable tests at feed salinity larger than 1.8 M were not available. In standard AGMD configuration, the values of the membrane leak ratio were always less than 0.024%. The fact that larger leak appeared for the high salinity feeds could indicate the presence of higher crystallization fouling, but as discussed before this would have a non-reversible effect which is not observed. For elucidating this matter, more detailed studies are needed analyzing the membranes behaviour with high salinity, something that can only be done to the full extent at a laboratory scale where membranes can be isolated and analyzed completely. Another interesting issue shown in Figure 6 that requires further investigation is that the membrane leak ratio was larger for the AS24 module than for the AS7 in V-AGMD (as expected by the larger membrane surface, as discussed in Section 3.2) but not in AGMD. When normalizing by the membrane surface area (Figure 7), however, it was confirmed that the AS7 module is actually the one with the higher leak. This suggests a difference in the membrane quality between the two modules used in these experiments (perhaps a larger ratio of pinholes per area in the AS7), or some other unaccounted effect influencing both modules differently. To assess the significance of this difference, an investigation comparing more modules of the same kind (AS7 and AS24) would be needed.

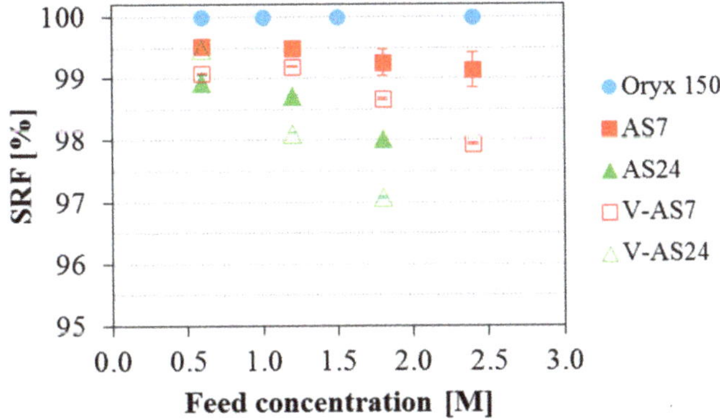

Figure 5. SRF obtained with the Oryx 150, the AS7 and the AS24 modules (in AGMD and V-AGMD configurations) for different initial feed concentrations. T_{hot} = 80 °C, T_{cold} = 25 °C and F = 1100 L·h^{-1} (AS modules) and 600 L·h^{-1} (Oryx 150).

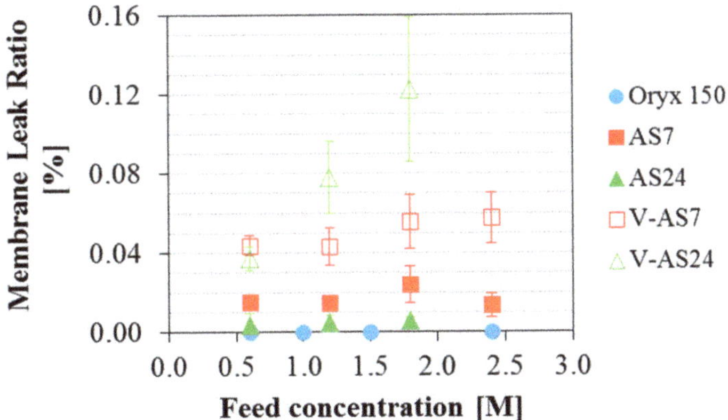

Figure 6. Membrane Leak Ratio obtained with the Oryx 150, the AS7 and the AS24 modules (in AGMD and V-AGMD configuration) for different initial feed concentrations. T_{hot} = 80 °C; T_{cold} = 25 °C; F = 1100 L·h^{-1} (AS modules) and 600 L·h^{-1} (Oryx 150).

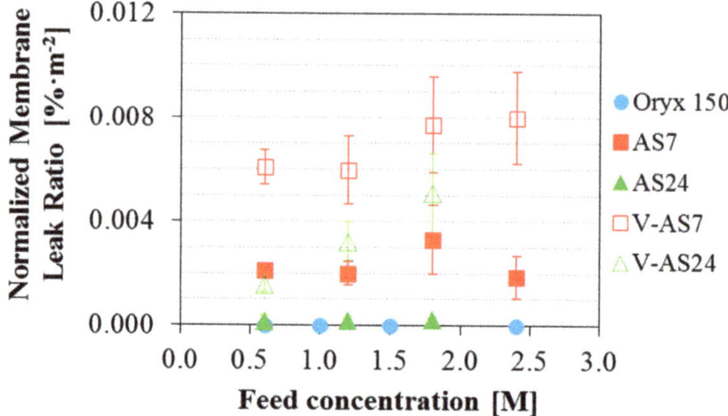

Figure 7. Normalized Membrane Leak Ratio obtained with the Oryx 150, the AS7 and the AS24 modules (in AGMD and V-AGMD configuration) for different initial feed concentrations. T_{hot} = 80 °C; T_{cold} = 25 °C; F = 1100 $L \cdot h^{-1}$ (AS modules) and 600 $L \cdot h^{-1}$ (Oryx 150).

4. Conclusions

Based on the results obtained from this work, it can be stated that MD demonstrates technological robustness for desalinating feed solutions with seawater concentration, because commercial modules analyzed in this study produced permeate of very good quality (SRF greater than 99%) during the entire period of testing (a large number of experiments performed along more than 300 h of operation), under very different operating conditions of T_{hot}, T_{cold} and F. However, intermittent operation generated a poor permeate quality when the process was restarted. This can be because liquid feed passes through possible membrane pinholes while no production of permeate occurs. When the operation is initialized, the permeate production dilutes the initial conductivity because the vapor flow is several orders of magnitude greater than the feed leak through the membrane. When the module was kept with demineralized water after cleaning, permeate had a very high quality from the beginning. The recovery of the permeate quality discards crystallization fouling, which would cause an irreversible deterioration of the membrane hydrophobicity. Permeate quality decreased with the increase of feed salinity, especially in the case of the Aquastill modules. The PTFE membrane used by the SolarSpring module showed much better retention than the LDPE membrane used by Aquastill, although in the worst case, the maximum leak through the membrane in the latter was never more than 0.12% of the feed. A conclusive reason cannot be given for this worsening of permeate quality with salinity. The fact that the membrane leak for the AS7 module was slightly larger than for the AS24 module, when their membrane is supposedly the same kind, suggests that other effects are present or the membranes properties can suffer some variation in the performance. More studies of these modules and the membranes behavior with high salinity feeds are needed to investigate this.

Author Contributions: Conceptualization, G.Z.; resources, G.Z.; formal analysis, A.R. and J.A.; investigation, A.R. and J.A.; validation, A.R. and J.A.; supervision, G.Z.; visualization, A.R., J.A. and G.Z.; writing—original draft preparation, A.R. and J.A.; writing—review and editing, G.Z and A.R.

Funding: This research received no external funding.

Acknowledgments: The authors want to acknowledge Rebecca Schwantes (SolarSpring), Bart Nelemans (Aquastill) and Jan Henk Hanemaaijer (i3 Innovative Technologies) for technical discussions.

Conflicts of Interest: The authors declare no conflict of interest.

Acronyms

AGMD	Air gap membrane distillation
CSG	Coal seam gas
DIC	Dissolved inorganic carbon
DTC	Dissolved total carbon
GOR	Gained output ratio
LDPE	Low density polyethylene
MD	Membrane distillation
MVC	Mechanical vapour compression
PP	Polypropylene
PTFE	Polytetrafluoroethylene
RO	Reverse osmosis
SRF	Salt rejection factor
V-AGMD	Vacuum-enhanced air gap membrane distillation
WHO	VWorld health organization

Nomenclature

C	Concentration ($g \cdot L^{-1}$)
EC	Electrical conductivity ($mS \cdot cm^{-1}$)
F	Feed flow rate ($L \cdot h^{-1}$)
T	Temperature (°C)
S	Salinity (M)

Subscripts

cold	Cold channel
f	Feed
hot	Hot channel
p	Permeate

Appendix A

The exact composition of the marine salts used in this work is not available. Typically, marine salts have a proportion of NaCl close to 90% and traces of other salts (gypsum, bischofite, langbeinite, dolomite, calcite, quartz and sylvine) [41].

Conversion from electrical conductivity (EC), in $mS \cdot cm^{-1}$, to concentration (c), in $g \cdot L^{-1}$, was made using Equation A1, derived from data tabulated for NaCl aqueous solutions at 20 °C [42,43], as presented in Figure A1.

$$C = a_1 \cdot EC + a_2 \cdot EC^2 + a_3 \cdot EC^3 + a_4 \cdot EC^4 \tag{A1}$$

where $a_1 = 0.5135$; $a_2 = 7.3802 \times 10^{-3}$; $a_3 = -6.5443 \times 10^{-5}$; $a_4 = 2.1907 \times 10^{-7}$. The maximum error of the correlation is 2.8% (0–300 $g \cdot L^{-1}$), and $R^2 = 0.9994$.

The correlation was chosen for NaCl aqueous solutions because that for seawater was not available for high salt concentrations. A comparison of this correlation with that for seawater resulted in errors lower than 4%. Since the concentration of NaCl in marine salts is larger than in seawater, it is expected that the error will be lower.

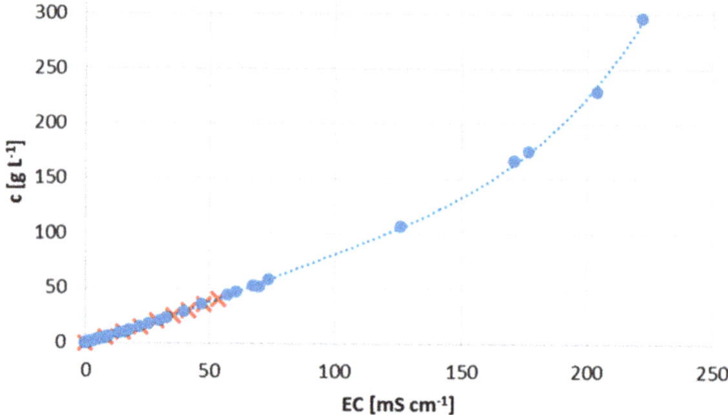

Figure A1. Salt concentration as a function of electrical conductivity. Blue dots: NaCl aqueous solution. Red crosses: seawater.

References

1. Shatat, M.; Worall, M.; Riffat, S. Opportunities for solar water desalination worldwide: Review. *Sustain. Cities Soc.* **2013**, *9*, 67–80. [CrossRef]
2. Sarai Atab, M.; Smallbone, A.J.; Roskilly, A.D. An operational and economic study of a reverse osmosis desalination system for potable water and land irrigation. *Desalination* **2016**, *397*, 174–184. [CrossRef]
3. Monnot, M.; Ngvyên, H.T.K.; Laborie, S.; Cabassud, C. Seawater reverse osmosis desalination plant at community-scale: Role of an innovative pretreatment on process performances and intensification. *Chem. Eng. Process. Process Intensif.* **2017**, *113*, 42–55. [CrossRef]
4. Shalaby, S.M. Reverse Osmosis desalination powered by photovoltaic and solar Rankine cycle power systems: A review. *Renew. Sustain. Energy Rev.* **2017**, *73*, 789–797. [CrossRef]
5. Roberts, D.A.; Johnston, E.L.; Knott, N.A. Impacts of desalination plant discharges on the marine environment. A critical review of published studies. *Water Res.* **2010**, *44*, 5117–5128. [CrossRef]
6. Nakoa, K.; Rahaoui, K.; Date, A.; Akbarzadeh, A. Sustainable zero liquid discharge desalination (SZLDD). *Sol. Energy* **2016**, *135*, 337–347. [CrossRef]
7. Macedonio, F.; Drioli, E. Zero Liquid Discharge in Desalination. In *Sustainable Membrane Technology for Water and Wastewater Treatment. Green Chemistry and Sustainable Technology*; Figoli, A., Criscuoli, A., Eds.; Springer: Singapore, 2017; pp. 221–241.
8. Ji, X.; Curcio, E.; Al Obaidini, S.; Di Profio, G.; Fontananova, E.; Drioli, E. Membrane Distillation—Crystallization of seawater reverse osmosis brines. *Sep. Purif. Technol.* **2010**, *71*, 76–82. [CrossRef]
9. Drioli, E.; Curcio, E. Progress in membrane crystallization. *Curr. Opin. Chem. Eng.* **2012**, *1*, 178–182. [CrossRef]
10. Quist-Jensen, C.A.; Macedonio, F.; Drioli, E. Membrane crystallization for salts recovery from brine—An experimental and theoretical analysis. *Desalin. Water Treat.* **2015**, *57*, 7593–7603. [CrossRef]
11. Quist-Jensen, C.A.; Macedonio, F.; Drioli, E. Integrated Membrane Desalination Systems with Membrane Crystallization Units for Resource Recovery: A new approach for mining from the sea. *Crystals* **2016**, *6*, 36–48. [CrossRef]
12. Ruiz Salmón, I.; Luis, P. Membrane crystallization via membrane distillation. *Chem. Eng. Process. Process Intensif.* **2018**, *123*, 258–271. [CrossRef]
13. Giwa, A.; Dufour, V.; Al Marzooqi, F.; Al Kaabi, M.; Hasan, S.W. Brine management methods: Recent innovations and current status. *Desalination* **2017**, *407*, 1–23. [CrossRef]
14. Tong, T.; Elimelech, M. The Global Rise of Zero Liquid Discharge for Wastewater Management: Drivers, Technologies, and Future Directions. *Environ. Sci. Technol.* **2016**, *50*, 6846–6855. [CrossRef] [PubMed]

15. Xu, J.; Singh, Y.B.; Amy, G.L.; Ghaffour, N. Effect of operating parameters and membrane characteristics on air gap membrane distillation performance for the treatment of highly saline water. *J. Membr. Sci.* **2016**, *512*, 73–82. [CrossRef]
16. Alkhudhiri, A.; Hilal, N. Air gap membrane distillation: A detailed study of high saline solution. *Desalination* **2017**, *403*, 179–186. [CrossRef]
17. Li, J.; Guan, Y.; Cheng, F.; Liu, Y. Treatment of high salinity brines by direct contact membrane distillation: Effect of membrane characteristics and salinity. *Chemosphere* **2015**, *140*, 143–149. [CrossRef] [PubMed]
18. Eykens, L.; Hitsov, I.; De Sitter, K.; Dotremont, C.; Pinoy, L.; Nopens, I.; Van der Bruggen, B. Influence of membrane thickness and process conditions on direct contact membrane distillation at different salinities. *J. Membr. Sci.* **2016**, *498*, 353–364. [CrossRef]
19. Safavi, M.; Mohammadi, T. High-salinity water desalination using VMD. *Chem. Eng. J.* **2009**, *149*, 191–195. [CrossRef]
20. Adham, S.; Hussain, A.; Minier Matar, J.; Dores, R.; Janson, A. Application of Membrane Distillation for desalting brines from thermal desalination plants. *Desalination* **2013**, *314*, 101–108. [CrossRef]
21. Naidu, G.; Jeong, S.; Choi, Y.; Vigneswaran, S. Membrane distillation for wastewater reverse osmosis concentrate treatment with water reuse potential. *J. Membr. Sci.* **2017**, *524*, 565–575. [CrossRef]
22. Naidu, G.; Choi, Y.; Jeong, S.; Hwang, T.M.; Vigneswaran, S.J. Experiments and modeling of a vacuum membrane distillation for high saline water. *Ind. Eng. Chem.* **2014**, *20*, 2174–2183. [CrossRef]
23. Gryta, M. Concentration of NaCl solution by membrane distillation integrated with crystallization. *Sep. Sci. Technol.* **2002**, *37*, 3535–3558. [CrossRef]
24. Tun, C.M.; Fane, A.G.; Matheickal, J.T.; Sheikholeslami, R. Membrane distillation crystallization of concentrated salts-flux and crystal formation. *J. Membr. Sci.* **2005**, *257*, 144–155. [CrossRef]
25. Drioli, E.; Ali, A.; Macedonio, F. Membrane distillation: Recent developments and perspectives. *Desalination* **2015**, *356*, 56–84. [CrossRef]
26. Minier-Matar, J.; Hussain, A.; Janson, A.; Benyahia, F.; Adham, S. Field evaluation ofmembrane distillation technologies for desalination of highly saline brines. *Desalination* **2014**, *351*, 101–108. [CrossRef]
27. Duong, H.C.; Chivas, A.R.; Nelemans, B.; Duke, M.; Gray, S.; Cath, T.Y.; Nghiem, L.D. Treatment of RO brine from CSG produced water by spiral-wound air gap membrane distillation—A pilot study. *Desalination* **2015**, *366*, 121–129. [CrossRef]
28. Winter, D.; Koschikowski, J.; Wieghaus, M. Desalination using membrane distillation: Experimental studies on full scale spiral wound module. *J. Membr. Sci.* **2011**, *375*, 104–112. [CrossRef]
29. Ruiz-Aguirre, A.; Alarcón-Padilla, D.-C.; Zaragoza, G. Productivity analysis of two spiral-wound membrane distillation prototypes coupled with solar energy. *Desalin. Water Treat.* **2015**, *55*, 2777–2785. [CrossRef]
30. Schwantes, R.; Bauer, L.; Chavan, K.; Dücker, D.; Felsmann, C.; Pfafferott, J. Air gap membrane distillation for hypersaline brine concentration: Operational analysis of a full-scale module–New strategies for wetting mitigation. *Desalination* **2018**, *444*, 13–25. [CrossRef]
31. Ruiz-Aguirre, A.; Andrés-Mañas, J.A.; Fernández-Sevilla, J.M.; Zaragoza, G. Modeling and optimization of a commercial permeate gap spiral wound membrane distillation module for seawater desalination. *Desalination* **2017**, *419*, 160–168. [CrossRef]
32. Ruiz-Aguirre, A.; Andrés-Mañas, J.A.; Fernández-Sevilla, J.M.; Zaragoza, G. Experimental characterization and optimization of multi-channel spiral wound air gap membrane distillation modules for seawater desalination. *Sep. Purif. Technol.* **2018**, *205*, 212–222. [CrossRef]
33. Gil, J.D.; Ruiz-Aguirre, A.; Roca, L.; Zaragoza, G.; Berenguel, M. Prediction models to analyse the performance of a commercial-scale membrane distillation unit for desalting brines from RO plants. *Desalination* **2018**, *445*, 15–28. [CrossRef]
34. World Health Organization. *Guidelines for Drinking-Water Quality*, 4th ed.; WHO Press: Geneva, Switzerland, 2011; p. 223.
35. Guillén-Burrieza, E.; Zaragoza, G.; Miralles-Cuevas, S.; Blanco, J. Experimental evaluation of two pilot-scale membrane distillation modules used for solar desalination. *J. Membr. Sci.* **2012**, *409*, 264–275. [CrossRef]
36. Sanmartino, J.A.; Khayet, M.; García-Payo, M.C.; El Bakouri, H.; Riaza, A. Desalination and concentration of saline aqueous solutions up to supersaturation by air gap membrane distillation and crystallization fouling. *Desalination* **2016**, *393*, 39–51. [CrossRef]

37. Nguyen, Q.M.; Lee, S. Fouling analysis and control in a DCMD process for SWRO brine. *Desalination* **2015**, *367*, 21–27. [CrossRef]
38. Essalhi, M.; Khayet, M. Application of a porous composite hydrophobic/hydrophilic membrane in desalination by air gap and liquid gap membrane distillation: A comparative study. *Sep. Purif. Technol.* **2014**, *133*, 176–186. [CrossRef]
39. Peñate, B.; García-Rodríguez, L. Current trends and future prospects in the design of seawater reverse osmosis desalination technology. *Desalination* **2012**, *284*, 1–8. [CrossRef]
40. Schwantes, R.; Cipollina, A.; Gross, F.; Koschikowski, J.; Pfeifle, D.; Rolletschek, M.; Subiela, V. Membrane distillation: Solar and waste heat driven demonstration plants for desalination. *Desalination* **2013**, *323*, 93–106. [CrossRef]
41. Kovač, N.; Glavaš, N.; Donelec, M.; Šmuc, N.R.; Šlejkovec, Z. Chemical Composition of Natural Sea Salt from the Sečovlje Salina (Gulf of Trieste, northern Adriatic). *Acta Chim. Slov.* **2013**, *60*, 706–714.
42. Lide, D.R. *CRC Handbook of Chemistry and Physics*, 96th ed.; CRC Press: Boca Raton, FL, USA, 2015.
43. Washburn, E.W.; West, C.J. *International Critical Tables of Numerical Data, Physics, Chemistry and Technology*; McGraw Hill: New York, NY, USA, 1986.

MDPI
St. Alban-Anlage 66
4052 Basel
Switzerland
Tel. +41 61 683 77 34
Fax +41 61 302 89 18
www.mdpi.com

Membranes Editorial Office
E-mail: membranes@mdpi.com
www.mdpi.com/journal/membranes

www.ingramcontent.com/pod-product-compliance
Lightning Source LLC
LaVergne TN
LVHW070926170726
843515LV00005B/881

* 9 7 8 3 0 3 6 5 1 2 1 1 2 *